Mario Zaiss

The Role of Regulatory T Cells in Bone Homeostasis

AF010305

Mario Zaiss

The Role of Regulatory T Cells in Bone Homeostasis

Linking Cells from the Immune System with Bone Cells

Südwestdeutscher Verlag für Hochschulschriften

Impressum / Imprint
Bibliografische Information der Deutschen Nationalbibliothek: Die Deutsche Nationalbibliothek verzeichnet diese Publikation in der Deutschen Nationalbibliografie; detaillierte bibliografische Daten sind im Internet über http://dnb.d-nb.de abrufbar.

Alle in diesem Buch genannten Marken und Produktnamen unterliegen warenzeichen-, marken- oder patentrechtlichem Schutz bzw. sind Warenzeichen oder eingetragene Warenzeichen der jeweiligen Inhaber. Die Wiedergabe von Marken, Produktnamen, Gebrauchsnamen, Handelsnamen, Warenbezeichnungen u.s.w. in diesem Werk berechtigt auch ohne besondere Kennzeichnung nicht zu der Annahme, dass solche Namen im Sinne der Warenzeichen- und Markenschutzgesetzgebung als frei zu betrachten wären und daher von jedermann benutzt werden dürften.

Bibliographic information published by the Deutsche Nationalbibliothek: The Deutsche Nationalbibliothek lists this publication in the Deutsche Nationalbibliografie; detailed bibliographic data are available in the Internet at http://dnb.d-nb.de.

Any brand names and product names mentioned in this book are subject to trademark, brand or patent protection and are trademarks or registered trademarks of their respective holders. The use of brand names, product names, common names, trade names, product descriptions etc. even without a particular marking in this work is in no way to be construed to mean that such names may be regarded as unrestricted in respect of trademark and brand protection legislation and could thus be used by anyone.

Verlag / Publisher:
Südwestdeutscher Verlag für Hochschulschriften
ist ein Imprint der / is a trademark of
OmniScriptum GmbH & Co. KG
Heinrich-Böcking-Str. 6-8, 66121 Saarbrücken, Deutschland / Germany
Email: info@svh-verlag.de

Herstellung: siehe letzte Seite /
Printed at: see last page
ISBN: 978-3-8381-1583-2

Zugl. / Approved by: Erlangen, Universität Erlangen-Nürnberg, Diss., 2009

Copyright © 2010 OmniScriptum GmbH & Co. KG
Alle Rechte vorbehalten. / All rights reserved. Saarbrücken 2010

The Role of Regulatory T Cells in Bone Homeostasis

Table of Contents

Table of Contents

Abstract   5
Zusammenfassung (german summary)   7

1. Introduction   9
- 1.1. Bone: Function and structure   9
- 1.1 1. Bone development   9
- 1.1.2. Bone cells   10
- 1.1.3. Bone remodeling   14
- 1.2. T cells   16
- 1.2.1. General properties of the immune response   16
- 1.2.2. $CD4^+$ T cells   16
- 1.2.3. Regulatory T cells   18
- 1.3. Osteoimmunology: Crosstalk between bone and the immune system   19
- 1.3.1. Osteoblasts and immune cells   19
- 1.3.2. Osteoclasts and immune cells   20

2. Material and Methods   23
- 2.1. Materials   23
- 2.2. Methods   27

3. Results   35
- 3.1. Suppressive effect of Tregs on osteoclastogenesis *in vitro*   35
- 3.1.1. Isolation and characterization of Tregs and osteoclast precursors   35
- 3.1.2. Regulatory T cells suppress osteoclast formation   36
- 3.1.3. Direct cell-contact is essential for suppression   42
- 3.1.4. CTLA-4 interaction to CD80/86 is mediating the suppression of osteoclast differentiation by Tregs   44
- 3.2. The role of Tregs in bone homeostasis   46
- 3.2.1. Increased bone mass in *foxp3*-transgenic (*foxp3*tg) mice   46
- 3.3. Tregs and bone pathology   53
- 3.3.1. *Foxp3*tg mice are protected from ovariectomy-induced bone loss   53

3.3.2.	*Foxp3*tg bone marrow protects from inflammation-induced bone loss	55
3.3.3.	SupCD28Mab ameliorates TNF-induced arthritis and systemic bone loss	59
3.4.	**Mechanism of osteoclast suppression by Tregs**	**61**
3.4.1.	Increased osteoclast numbers in $CD80/86^{-/-}$ mice	61
3.4.2.	IDO is essential for CTLA-4-mediated osteoclast suppression	61
3.4.3.	Increased bone density in $RAG1^{-/-}$ mice after adoptive Treg transfer	65

4. Discussion   69
4.1.	**Tregs as potential negative regulators of bone remodeling**	**69**
4.2.	**Mechanisms of osteoclast suppression by Tregs**	**71**
4.3.	**Tregs in bone diseases and normal bone homeostasis**	**74**

5. Concluding Remarks   79

7. References   81

Abstract

The objective of this Ph.D. thesis was to investigate the role of regulatory T cells (Tregs) in physiological and pathological bone remodeling.
We first *in vitro* characterized the suppressive effects of Tregs on the differentiation of the bone resorptive cells or osteoclasts. The effect was found to be directly mediated and to require cell to cell contact between the monocytic precursors of osteoclasts and Tregs. In addition, Tregs mediated expression of TGFβ, IL-4 and IL-10 contributed, but was not essential, to the inhibitory effect on osteoclastogenesis. The co-stimulatory receptor CTLA-4 expressed on the surface of Tregs was the main mediator of their suppressive effect that was abolished in monocytes lacking both the CTLA-4 ligands CD80 and CD86. These *in vitro* data showing that $CD4^+CD25^+Foxp3^+$ Tregs suppress osteoclast formation, prompted us to *in vivo* investigate the role of Tregs in bone homeostasis. We therefore analyzed bone parameters of mice with increased Treg numbers (*foxp3*-transgenic mice). Indeed *foxp3*tg mice developed high bone mass and were protected from ovariectomy-induced osteoporosis; inflammatory osteopenia and arthritic bone destruction, whereas *foxp3*-deficiency enhanced bone loss. Tregs directly protect against bone loss since a significant increased bone mass was observed following adoptive Tregs transfer in RAG1 deficient mice ($RAG1^{-/-}$). The positive skeletal effects of Tregs were entirely mediated through inhibition of bone resorption by osteoclasts. Binding of Tregs to CD80/CD86 on osteoclast precursors was essential for inhibiting osteoclast differentiation *in vivo* as shown by the osteopenic phenotype that developed in CD80/86 deficient mice ($CD80/86^{-/-}$) due to increased osteoclast differentiation. Mechanistically, engagement of CD80/86 by Tregs through CTLA-4 induced IDO expression in osteoclast precursors and thereby increased tryptophan catabolism and apoptosis. These effects were not found in CD80/86 deficient osteoclast precursors.
In summary, the presented *in vitro* and *in vivo* results demonstrate that $Foxp3^+$ Tregs can actively and directly control bone resorption during physiological and pathological bone remodeling and thus preserve bone mass. Consequently, immune regulation is actively protecting bone mass.

Abstract

Zusammenfassung

Ziel dieser Doktorarbeit war es, zu klären ob und wie regulatorische T Zellen (Tregs) die Osteoklasten Differenzierung *in vitro* sowie den Knochenstoffwechsel *in vivo* beeinflussen können. Diese Fragestellungen sollten nicht nur einen tieferen Einblick in die Interaktion von Immunsystem und Knochen erzielen, sondern auch dazu dienen, Strategien zu entdecken, wie das Immunsystem den Knochenabbau verhindert.

Dazu wurden $CD4^+CD25^+Foxp3^+$ T Zellen (Tregs) isoliert und mit $CD11b^+$ Osteoklastenvorläuferzellen kultiviert. Die Osteoklasten Differenzierung wurde mittels einer Tartrat-resistenten sauren Phosphatase (TRAP) Färbung nach 4-6 Tagen in Kultur untersucht. Zudem wurde die Fähigkeit der Osteoklasten zur Knochenresoprtion mittels Resorptionassay auf Osteologic ^{TM} Platten analysiert, Transwell-Kulturexperimente durchgeführt und Zytokine blockiert, um die Mechanismen der Interaktion zwischen Tregs und Osteoklasten aufzuklären. Wir konnten zeigen, dass regulatorische $CD4^+CD25^+Foxp3^+$ T Zellen im Gegensatz zu $CD4^+CD25^-$, $Foxp3^-$ T Zellen die Osteoklastendifferenzierung durch M-CSF und RANKL dosisabhängig blockieren. Deren resorptive Aktivität war bis zu 80% reduziert sobald Tregs zu Osteoklasten Vorläuferzellen gegeben wurden. Die Blockade der Osteoklastendifferenzierung basierte nicht auf einem Ungleichgewicht zwischen RANKL und OPG Verhältnis, sondern war wesentlich von einem direkten Zell-Zell Kontakt zwischen CTLA-4 auf Tregs und CD80/86 auf Osteoklastenvorläuferzellen abhängig. Die Expression von TGFβ, IL-4 und IL-10 durch Tregs, hatte einen zusätzlichen blockierenden Effekt auf die Osteoklastendiffernzierung, war aber nicht essentiell für die inhibitorische Wirkung von Tregs auf die Osteoklastogenese verantwortlich.

Diese ersten *in vitro* Daten zeigen, dass $CD4^+CD25^+Foxp3^+$ Tregs die Differenzierung von Osteoklasten aus Osteoklastenvorläuferzellen hemmen können und zeigen eine neue regulierende Beziehung zwischen dem Immunsystem und dem Knochenstoffwechsel.

In den folgenden *in vivo* Versuchen stellten wir uns die Frage ob diese regulierende Beziehung zwischen dem Immunsystem und dem Knochenstoffwechsel auch *in vivo* nachweisbar ist. Dazu konnten wir zeigen, dass Mäuse mit einem erhöhten Anteil an Tregs (*foxp3*-transgene Mäuse) eine signifikant erhöhte Knochendichte aufweisen.

Zudem waren *foxp3*-transgene Mäuse sowohl von der postmenopausalen Osteoporose als auch von inflammatorischer Osteopenie und arthritischer Knochendestruktion geschützt. Tregs erhöhten zudem die Knochenmasse in Lymphozyten-defiziente $RAG1^{-/-}$ Mäuse ($RAG1^{-/-}$), wobei nach adoptiven Transfer von Treg eine signifikant höhere Knochendichte zu beobachten war.

Die beschriebenen positiven Effekte der Tregs auf die systemische Knochendichte wurden durch die Hemmung der Knochenresoption durch Osteoklasten verursacht. Der direkte Kontakt von Tregs mit CD80/86 auf Osteoklastenverläuferzellen war essentiell für die Hemmung der Knochenresoption durch Osteoklasten. In Übereinstimmung damit zeigen CD80/86 knock-out Mäuse (*CD80/86-/-*) einen osteoporotischen Phänotyp basierend auf einer erhöhten Anzahl an Osteoklasten. Die Binding von CD80/86 durch CTLA-4 auf Tregs induziert in Osteoklastenvorläuferzellen die Expression von IDO und somit einen verstärkten Tryptophan Stoffwechsel was dazu führt das Osteoklastenvorläuferzellen vermehrt apoptotisch werden und sich somit nicht in reife knochenresorbierende Osteoklasten differenzieren können. Osteoklastenvorläuferzellen aus *CD80/86-/-* Mäusen zeigen keine Hochregulierung der IDO mRNA Expression und werden nach Kontakt mit CTLA-4 Ig oder Tregs auch nicht apoptotisch.

Zusammenfassend kann man sagen dass die hier präsentierten *in vitro* sowie *in vivo* Daten zeigen, dass Tregs direkt die Knochenresorption sowohl im physiologischen als auch im pathologischen Knochenstoffwechsel kontrollieren und somit eine neue wichtige Verbindung zwischen dem Immunsystem und dem Knochenstoffwechsel darstellen.

1. Introduction

1.1. Bone: Function and structure

Bone is not an inert, static material. It is a highly organized, living tissue that constitutes the skeleton. Bones have many functions: they protect the internal organs, they support the body structures, and bones serve as an attachment site for muscles allowing locomotion. Most importantly bones serve as an appropriate cavity for the bone marrow and thus support hematopoiesis and act as a reservoir for inorganic ions being responsible for maintenance of calcium homeostasis and rapidly mobilization mineral stores on metabolic demand. Bone is composed of cells and extracellular matrix, the latter being further subdivided into an organic and inorganic part. The organic matrix is mainly constituted of type I collagen (approximately 95%), as well as other types of collagens, non-collagenous proteins and proteoglycans, whereas the inorganic matrix predominantly contains calcium and phosphorus, appearing as hydroxyapatite crystals deposited into the collagenous matrix.

1.1.1. Bone development

Bone development occurs along two pathways, namely the endochondral bone and the membranous bone development. Endochondral bone, which includes the long bones and vertebrae, develops in the embryo as a cartilaginous template. This cartilaginous template is defined by mesenchymal cell condensations. In the centre of these condensations, mesenchymal cells differentiate into chondrocytes which express specific markers like collagen II (the main constituent of cartilage) and aggrecan. The chondrocytes differentiate further into pre-hypertrophic and hypertrophic chondrocytes which can be distinguished by morphology and altered expression level of collagen II and increased level of collagen X. While pre-hypertrophic chondrocytes express a very low level of collagen II, hypertrophic chondrocytes express instead collagen X. At the perichondrium, the periphery of the condensation, the mesenchymal cells differentiate into osteoblasts which build the bone collar that surrounds the newly formed cartilaginous scaffold. Subsequently, blood vessels penetrate into the cartilaginous scaffold leading to endochondral ossification. Blood vessels permit the invasion by osteoclasts that resorb most of the newly mineralized cartilage matrix. Additionally osteoblast progenitors which are brought in from the bone collar replace the cartilage matrix by deposing collagen I, the main component of the bone matrix.

Endochondral bone is then organized into cortical and trabecular bone. Cortical bone forms the outer surface of endochondral bones and provides the structural integrity for many of the long bones. It forms up to 80% of the skeleton and is typically dense bone with a well-organized pattern of collagen fibrils that are aligned along stress lines to provide bone with maximum strength (1, 2). Trabecular bone is thinner and less well organized, and it is primarily found traversing the bone marrow space. However, in vertebrae, trabecular bone is the main mechanical provider of structural integrity. A major function of trabecular bone is to provide a large surface area for metabolic processes. Bone turnover, which consists of bone resorption and its replacement with new bone, occurs much more rapidly in trabecular bone than in cortical bone.

In contrast to long bones and their endochondral developmental pathway, most bones of the skull and flat bones are generated by a process called membranous ossification. This type of bones is directly formed from the condensation of mesenchymal cells directly differentiating into osteoblasts which in turn depose the mineralized bone matrix.

1.1.2. Bone cells

Osteoclasts

The osteoclast derives from hematopoietic stem cells and is a member of the monocyte/macrophage family. It is a highly specialized multinucleated macrophage which is distinguished by its polarization (induced by contact with bone) and its expression of specific proteins such as cathepsin K, a highly specific intracellular organization and its capacity to actively resorb bone. The most notable feature of the polarized osteoclast is its ruffled membrane. This structure, which constitutes the resorptive organelle of the cell, consists of a unique villous-like complex of the plasma membrane that is juxtaposed to bone and contains numerous 'spike-like' vacuolar proton pumps (H^+ATPases). Formation of the osteoclast ruffled membrane is dependent on contact with bone and is only apparent when the cell degrades skeletal matrix. Osteoclast attachment to bone surface is dependent on several integrins expressed in osteoclasts that possesses specific amino acid sequences within proteins at the surface of the bone matrix (3). After attachment to the bone matrix, αVβ3 integrin binding activates cytoskeletal reorganization within the osteoclast. Integrin signaling and subsequent podosome formation, that are necessary for the movement across the bone surface, is dependent on a number of adhesion kinases including the proto-oncogene Src (4).

Two bone disorders depend on osteoclast activity, (i) osteopetrosis with decreased osteoclast activity and resulting increased bone mass and (ii) osteoporosis with increased osteoclast activity and resulting decreased bone mass.

Osteoclast differentiation requires two essential molecules, (i) the macrophage colony-stimulating factor (M-CSF) and (ii) receptor for activation of nuclear factor kappa B (NF-κB) (RANK) ligand (RANKL) (also known as OPGL, ODF and TRANCE). The binding of M-CSF to its receptor c-FMS on osteoclast precursors is important for osteoclast survival and proliferation (5). M-CSF can induce its own receptor expression, thereby forming an autocrine loop to amplify M-CSF-mediated signals, while it has also been reported to stimulate PU.18 (6). Activation of c-FMS by M-CSF is necessary for the proliferation and survival of macrophage/osteoclast progenitor cells. Binding of M-CSF to c-FMS signals via the PI3K and ERK, MAPK pathways. Mice with an inactivating mutation of M-CSF (*op/op* mice) lack macrophages and osteoclasts and develop osteopetrosis (7). In addition, M-CSF up-regulates the expression of RANK on osteoclast precursors (8). The second essential signal, RANKL stimulates the pool of M-CSF–expanded precursors to precede to the osteoclast phenotype. RANKL administration in mice results in increased bone resorption (9), while RANKL deficient mice fail to develop osteoclasts (1). Moreover, complete osteoclastogenesis can now be achieved *in vitro* with pure populations of monocytes exposed only to M-CSF and RANKL (2). Osteoprotegerin (OPG) is the soluble decoy receptor for RANKL and therefore competes with RANK for RANKL. This is highlighted by OPG knock out mice which are osteoporotic due to increased osteoclast differentiation (10, 11). Therefore the balance between RANKL and OPG is determinant for osteoclast differentiation *in vivo* and *in vitro* (12). The essential role of RANKL in osteoclast differentiation is demonstrated by the phenotype of mice in which this gene has been deleted. RANKL knockout mice (*RANKL$^{-/-}$*) exhibit severe osteopetrosis and defective tooth eruption associated with a complete absence of osteoclasts (1), irregular bone surfaces because of the absence of remodeling, abnormal growth plates with club-shaped long bones, and growth retardation at several skeletal site (13). In line with these observations in *RANKL$^{-/-}$* mice also RANK knockout mice (*RANK$^{-/-}$*) lack osteoclasts and have therefore a severe defect in bone remodeling and in the development of cartilaginous growth plates of endochondral bones (14). The essential molecules for osteoclast differentiation such as RANKL, RANK, and OPG are all members of the TNF-TNFR superfamily proteins (12, 15). Binding of RANKL to its receptor RANK starts the intracellular signal transduction among which TNF-receptor activating factor 6 (TRAF6) plays a critical role (16) and induce NF-*k*B, AP-1, and NFATc1 activation, which are all necessary transcription factors for osteoclast differentiation (17-19).

Osteoblasts

The osteoblast is a mononuclear cell of mesenchymal origin that is responsible for bone formation. Osteoblasts produce osteoid, the initially uncalcified bone substance, which is mainly composed of type I collagen. Osteoblasts are also responsible for mineralization of the osteoid matrix. The process of mineralization of bony tissue requires high concentrations of alkaline phosphatase in osteoblasts. Genetically, the osteoblast can be viewed as a specialized mesenchymal cell, since all genes expressed in mesenchymal cells are also expressed in osteoblasts. Two bone disorders depend on osteoblast activity, (i) osteopenia with decreased osteoblast activity and resulting in decreased bone mass and (ii) osteosclerosis with increased osteoblast activity resulting in increased bone mass.

Core binding factor alpha1 (Cbfa1) (also known as runt-related transcription factor 2 (Runx2)) (20), a transcription factor, as well as the genes OG1 and OG2 encoding for osteocalcin, a secreted molecule that inhibits osteoblast function (21) are osteoblast specific transcripts. This is highlighted by studies with Runx2-deficient (*runx2$^{-/-}$*) mice, which completely lack bone formation due to the absence of osteoblasts (22, 23). In addition, studies with osteocalcin-deficient mice (*osc^{m1}/osc^{m1}*) showed that these mice develop a phenotype marked by higher bone mass and improved bone quality (21). Beside these two osteoblast specific transcripts, osteoblast differentiation is regulated by many secreted factors including TGFβ, bone morphogenetic proteins (BMPs), insulin-like growth factors (IGFs), parathyroid hormone (PTH), PTH-related peptide, Indian hedgehog, retinoic acid and Wnts.

Osteocytes

In contrast to active cuboid- shaped osteoblasts, resting osteoblasts acquire a spindle shaped appearance. Resting osteoblasts also loose their ability to proliferate. As a result of the production of osteoid in the osteoblasts neighborhood, that finally undergoes mineralization, the osteoblast are incorporated into the bone matrix and become osteocytes. Osteocytes are the most abundant cells of the bone, and they form a network within the mineralized bone allowing metabolic exchange. Despite the complex organization of osteocytes network, the exact function remains unknown so far. It is likely that osteocytes respond to bone damage and enhance bone remodeling by recruiting osteoclasts to sites where bone remodeling is required. (24).

Chondrocytes

The chondrocyte is a mononuclear cell of mesenchymal origin that is directly linked to the process of endochondral ossification of the long bones and for regulation of longitudinal bone growth. Chondrocytes can be divided into non-hypertrophic and hypertrophic chondrocytes. Non-hypertrophic chondrocytes include the resting and proliferating chondrocytes situated at the reserve and proliferating cartilage zone at the bone epiphysis, respectively. That non-hypertrophic chondrocyte proliferation is crucial for longitudinal bone growth has been demonstrated by a mutation in the fibroblast growth factor receptor 3 (FGFR3), an essential receptor for chondrocyte proliferation resulting in the human disorder achondroplasia (ACH). The complete absence of non-hypertrophic chondrocytes in ACH leads to dwarfism phenotype in these patients (25). However, after proliferation, non-hypertrophic chondrocytes initiate their terminal differentiation which goes along with changes in the gene expression profile and a dramatic increase in cell volume. These swollen chondrocytes are now termed hypertrophic chondrocytes which mineralize their extracellular matrix and undergo apoptosis. This terminal differentiation step is controlled by the Wnt signaling pathway where Wnt5a and Wnt5b, out of the 19 known secreted members (26), seems to play a critical role (27). Additional factors driving the differentiation step are the parathyroid hormone-related peptide (PTHrP) as a negative and Indian hedgehog (Ihh) as a positive regulator of chondrocytes differentiation (28). The upregulation of the vascular endothelial growth factor (VEGF) in hypertrophic chondrocytes under the control of Cbfa1 (Runx2) is the limiting step to start vascularization and subsequent ossification of the extracellular matrix (29, 30) for final bone formation.

1.1.3. Bone remodeling

Bone remodeling is defined by the removal of mineralized bone by osteoclasts followed by the formation of new bone matrix by osteoblasts. This process can be divided into three consecutive phases, (i) resorption, when bone is digested by osteoclasts, (ii), reversal, when mononuclear cells appear on the bone surface, and (iii) formation, when osteoblast deposit new bone matrix. The process of bone remodeling is controlled by many factors regulating the bone cells at systemic or at local levels.

Regarding the systemic regulation of bone remodeling, parathyroid hormone (PTH), calcitriol (the active form of vitamin D3), insulin-like growth factors (IGFs), glucocorticoids, estrogens and molecules involved in the Wnt pathway are the main regulatory factors of systemic regulation of bone remodeling. PTH is essential for maintaining serum calcium levels by stimulating bone resorption through osteoclasts (31). Calcitriol supports the intestinal calcium and phosphorus absorption (32) and stimulates the secretion of osteocalcin by osteoblast which in turn enhances the calcium integration into the bone matrix (33). Growth hormone (GH), IGF-1 and IGF-2 enhances longitudinal bone growth by promoting chondrocyte hypertrophy (34). Glucocorticoids have both stimulatory and inhibitory effects on bone remodeling: On the one hand, they are essential for osteoblast differentiation and on the other hand they are capable of suppressing osteoblast activity (35). Estrogens, especially 17β-estradiol (E_2), also has strong systemic regulatory effects on bone remodeling as they are able to directly inhibit osteoclastic resorption by directly inducing apoptosis in theses cells (36). In addition, estrogens decrease the responsiveness of osteoclast precursors to RANKL stimulation, thereby preventing osteoclast differentiation (37). Mesenchymal precursor cells have the ability to differentiate into adipocytes, chondrocytes, or osteoblasts, depending on the signaling pathways influencing the surrounding milieu. Molecules of the Wnt pathway, especially Wnt10b encourage these cells to differentiate into skeletal precursors and prevent them from differentiating into pre-adipocytes. High levels of canonical Wnt signaling, as indicated by high levels of β-catenin, expressing Runx2, promote osteoblastogenesis at the expense of chondrocyte differentiation. In contrast, low levels of β-catenin along Sox9 lead to chondrocyte differentiation (38). Wnt signaling is modulated by several different families of secreted negative regulators. Among these, Dickkopf (DKK) is a family of cysteine-rich proteins comprising at least four different forms (DKK-1, DKK-2, DKK-3 and DKK-4). The best studied of these is DKK-1, which functions as a natural inhibitor of Wnt signaling (39, 40). When DKK-1 binds to the LPR5/6 receptor and a cell surface co-receptor, Kremen-1/2, it promotes internalization of the receptor complex and dampens the Wnt signal. Deletion of a single allele of DKK-1 increases bone mass in mice (41). Therefore, it was shown that the expression of DKK-1 in inflammatory and degenerative joint diseases inhibits bone formation by osteoblasts, and shift the balance in bone remodeling towards bone resorption by osteoclasts (42).

In regard to local regulation of bone remodeling, the already mentioned RANKL/RANK/OPG systems plays the most prominent role (43). Initiation of osteoclastogenesis largely depends on the direct local interaction between osteoclast precursor cells with cells from the osteoblast lineage. Osteoblasts are the local source of M-CSF and RANKL for osteoclast precursors. In addition, osteoblasts recruit osteoclast precursors to the site where bone remodeling is necessary by expressing the monocyte chemoattractant protein-1 (MCP-1, also known as CCL2) (44). Interestingly, the expression on osteoclast precursors of the ligand for MCP-1, MCP-1 receptor, is induced by RANKL, which is also secreted by osteoblasts (45). For direct bidirectional signaling between osteoclasts and osteoblasts are the Eph tyrosine kinase receptors, especially the EphB4 on osteoblasts, and its ligand epherinB2 on osteoclasts important. Reverse signaling through the epherinB2 ligand into osteoclasts suppresses their differentiation, while forward signaling through the EphB4 receptor into osteoblast precursors enhances their differentiation (46). It was speculated that since epherinB2 is expressed on mature osteoclasts and its ligand EphB4 on osteoblast precursors, that the epherinB2/EphB4 interaction starts the transition from bone resorption to bone formation (47). In addition for local regulatory mechanisms, several studies have shown a stimulating effect of IL-6, both on the bone resorptive capacity of osteoclasts (48) as well as on the generation on osteoblasts (49). Furthermore cytokines such as TNFα or IL-1β can regulate bone remodeling by enhancing RANKL expression in osteoblasts and thereby increasing bone resorption by osteoclasts (50).

1.2. T cells

1.2.1. General properties of the immune response

The immune system defends the host against infection. Innate immunity serves as a first line in defense. Adaptive immunity is based on clonal selection from a repertoire of lymphocytes with highly diverse antigen-specific receptors that enable the immune system to recognize any foreign antigen. Upon recognition of their specific antigen, naive lymphocytes become activated, start to proliferate and differentiate into antigen-specific effector cells. Effector lymphocytes are able to drive elimination of infections either by producing soluble molecules (antibodies (Ab), humoral immunity) or by activating other cells of the immune system (cellular immunity). Some effector cells become memory lymphocytes that can subsequently respond more rapidly, vigorously, and effectively to the antigen during later infections. In general, lymphocytes are divided into T lymphocytes (T cells) and B lymphocytes (B cells). B cells produce antibodies and are therefore the major constituents of humoral immunity. Activated B cells also provide signals to T cells, for example via the B7-family molecules that promote their continued activation.

A clear distinction between host- (self or auto) and foreign- (non-self) antigens is necessary for successful protection by the immune system from pathogens. T lymphocytes, which are able to recognize self-antigens (autoreactive T cells), are usually negatively selected in the thymus during their development and eliminated. This is a very effective selection process with only few T cells being able to escape the selection. Recognition of self-antigens by peripheral autoreactive T cells occurs frequently and is termed autoimmunity. The regulatory mechanisms in the periphery exerted by specific T cells (peripheral tolerance) are capable to prevent sustained immune responses to self-antigens and, thereby, the initiation of destructive autoimmune responses against host tissue, that might lead to autoimmune diseases.

1.2.2. CD4$^+$ T cells

T cells originate in the thymus from bone marrow precursors, which derive from the hematopoietic stem cell pool. Signals from the thymus lead to the migration and the seeding of the bone marrow precursors into the thymus. The early thymic progenitor cell is the intrathymic bone marrow derived T cell precursor (51). The intrathymic T cell precursor cells are the double-negative (DN) T cell precursors at stage 1 (DN1) and will undergo three consecutive stages, namely DN2, DN3, and DN4 before they acquire the T cell receptor (TCR) (52). When the immature T cells express the pre-TCR, they become CD4 and CD8 double positive before they rearrange the TCRα chain. Rearranging the TCRα chain leads to the positive or negative selection to naïve CD4 or CD8 single positive T cells that are exported to the periphery. For the effective host defense the differentiation of naive $CD4^+$ T cell precursors into the appropriate effector subset is critical. The differentiation decision is made predominantly by the cytokines in the microenvironment after activation and to some extent, by the strength of the interaction of the T cell antigen receptor (TCR) with the antigen (53). As known so far, $CD4^+$ T cells can differentiate into several different effector subsets, including T helper 1 (Th1) and T helper 2 (Th2) cells, the more recently defined Th17 cells, follicular helper T (Tfh) cells, and induced regulatory T cells (iTreg).

Th1 and Th2 lineages

Th1 cells have been identified by the expression of distinct cytokines such as IL-2, IFN-γ and lymphotoxin-α and are involved in cellular immunity against intracellular microorganisms, whereas Th2 cells mainly express IL-4, IL-5, IL-9 and IL-13 (54, 55). IL-12, produced by dendritic cells (DC), monocytes, macrophages and B cells as well as IFNγ, produced by both natural killer T (NK) cells and T cells, polarize towards Th1 differentiation through action of the signal transducer and activator transcription 4 (STAT4), STAT1, and T box transcription factor T-bet (56). The initiation of Th2 cell differentiation is dependent on TCR-mediated signaling, IL-33 and the IL-4 receptor mediated STAT6 activation and subsequent upregulation of GATA3 transcription (57).

Th17 cells

Th17 cells produce IL-17A (IL-17), IL-17F, and IL-22 and play important roles in protection against extracellular bacteria, fungi and microbes probably not well covered by Th1 or Th2 immunity that evolved to enhance clearance of intracellular pathogens and parasitic helminthes, respectively. Th17 cell differentiation requires retinoid-related orphan receptor (ROR)γt, a transcription factor that is induced by TGF-β in combination with the pro-inflammatory cytokines IL-6, IL-21, and IL-23. All these differencing stimuli activate STAT3 phosphorylation (58, 59).

1.2.3. Regulatory T cells

In 1995, Sakaguchi and colleges published a study where they reported that minor populations of CD4$^+$ T cells that co-express the IL-2 receptor α chain (CD25) were capable of suppressing immune responses in a variety of experimental autoimmune diseases (60). These cells were then known as regulatory T cells (Tregs). Recent studies highlighted the role of Tregs and could identify an essential transcriptional regulator for Treg differentiation and maintenance, the X chromosome encoded forkhead/winged-helix box P3 (Foxp3). Inactivating mutations in Foxp3 are causing IPEX (immune dysregulation, polyendocrinopathy, enteropathy, X-linked) syndrome, an autoimmune disorder in human patients also observed in the *scurfy* mutant phenotype in mice (61-64). Stable expression of Foxp3 is needed for the differentiation (65, 66), the suppressor function, and the proliferation (67) of Tregs. The loss of Foxp3 or its reduced expression in Tregs leads to the loss of the Tregs suppressive phenotype and their change into an effector T cell phenotype with expression of cytokines such as IL-2, IL-4, and IL-17 (68, 69).

The mechanism of the Tregs-mediated suppressive effect is still controversial. One possibility is that Tregs mediate suppression by blocking the IL-2 mRNA expression in responder T cells (70, 71). Another possible mechanism is the competitive consumption of IL-2 by Tregs and depriving responder cells from IL-2, which is essential for their proliferation (72). It might also be possible that cytokines play an important role in the suppressive mechanism. Although it has been shown that the suppressive effect on responder cell was missing when Tregs were separated form effector cells, it could not be excluded that cytokines play an additional role. A recently described cytokine of the IL-12 cytokine family, IL-35, would fit in this concept (73). In addition to IL-35, it is reported that galectin-1, a member of the β-galactoside binding proteins, may play a role in the responder cell suppression, since binding of galcetin-1 on responder cells mimics the effect of the binding of Tregs: (i) cell cycle arrest, (ii) apoptosis and (iii) decrease in proinflammatory cytokine secretion (74). Contrary to directly suppress responder T cells, several studies investigated the indirect suppression of responder T cells via antigen presenting cells (APC) as targets. Therefore, the interaction of cytotoxic T-lymphocyte antigen 4 (CTLA-4, or CD152), that is constitutively expressed on Tregs, with the co-stimulatory molecules CD80 and CD86 on APCs seems to be essential. The importance of CTLA-4 on Tregs is highlighted by a study, in which mice with a selective deletion of CTLA-4 in Tregs spontaneously developed a systemic autoimmunity (75).

Serra and colleges (76) demonstrated that Tregs are able to down regulate the expression of CD80/86 on dendritic cells (DC). This results in a limited capacity of DCs to activate naïve T cells via CD28 and thereby preventing proliferation of responder cells. In line with these observations, interaction of CTLA-4 with CD80/86 on DC stimulates the expression if indoleamine 2,3-dioxygenase (IDO), a molecule that induces the catabolism of tryptophan into pro-apoptotic metabolites such as kynurenine. These pro-apoptotic metabolites lead to the suppression of activation of naïve responder cells (77).

1.3. Osteoimmunology: Crosstalk between bone and the immune system

In the relatively new research field of osteoimmunology researchers investigate how bone influences immune and hematopoietic cells as well as how hematopoietic and immune cells influence bone. It is well known that the development of the immune system, at least in adult mammals, depends on the normal function of hematopoietic stem cells (HSCs), which reside adjacent to bone cells. On the other hand, the maturation of immune cells occurs in part in the bone marrow and depends on a suitable microenvironment by bone cells. Since during ontogenesis bone development precedes early immune system development it is unlikely that the immune system influences early skeletal formation. However, bone homeostasis and remodeling occur throughout life in vertebrates, which makes the interaction between bone and the immune system very likely. In addition, the anatomic structures of bone with its formation of cellular compartments in the bone marrow spaces principally allow immune and bone cells to interact and influence each other.

1.3.1. Osteoblasts and immune cells

Osteoblasts are involved in several aspects of osteoimmunology. The role of osteoblasts in hematopoietic stem cell (HSC) niches is creating increasing interest (78). The interaction of osteoblasts, which are a crucial component of the niches, on the trabecular bone surface with HSCs is mediated by molecular interactions involving N-cadherin and β-catenin (79). Additionally, osteoblasts are importantly influencing the reservoir of memory B cells and T cells in bone. Antibody producing B cells or plasma cells up-regulate CXCR4 and the ligand for CXCR4 is highly expressed by osteoblasts and stromal cells, facilitating migration of B cells and plasma cells to the bone marrow. Long lived memory CD8 T cells show special affinity for bone marrow cells that allow them migrate to the bone marrow as well (80). Another link between osteoblast and immune cells are cytokines expressed by lymphocytes that regulate osteoblast differentiation and activity. It is known that TNFα can inhibit osteoblast differentiation (81). IL-1 and IFNγ are able to inhibit collagen synthesis in osteoblasts whereas IL-4 and IL-13 are known chemoattractants for osteoblasts (82, 83).

1.3.2. Osteoclasts and immune cells

In 1972, the results published by Horton and colleges were the first hint that immune cells are capable of influencing osteoclast differentiation and activity (84). They observed that supernatants from cultured human blood leukocytes stimulate bone resorption by osteoclasts. Horton and colleges named these factors osteoclast activating factors (OAF), which were 13 years later identified as IL-1β (85). In 1999, Kong and colleges found that the essential factor for osteoclast differentiation, RANKL, is also expressed on T cells and enhances osteoclast differentiation, which suggested a tight link between bone and the immune system (86). In addition, several cytokines secreted by T cells can either stimulate osteoclast differentiation such as, IL-1, TNFα, IL-6, IL-11, IL-15, and recently IL-17 or that can block osteoclast differentiation, such as IL-4, IL-10, IL-13, IL-18 and GM-CSF. It seems that the effects of T cells on osteoclast differentiation are dependent on the balance between positive and negative factors expressed by T cells. Recently, several studies investigated the role of T cells on bone homeostasis, especially on osteoclast differentiation. It was found that activated CD3[+] T cells are capable in supporting osteoclast differentiation *in vitro* by surface expression of RANKL and that IL-1α and TGFβ further enhanced osteoclast differentiation (87). However, it was also published that IFNγ produced by activated CD5[+] T cells, a T cell accessory activation antigen (88), suppressed RANKL signaling through down-regulation of TRAF6 and therefore inhibits osteoclast differentiation (89). Through surface expression of RANKL, activated CD4[+] T cells can stimulate osteoclast differentiation *in vitro* only in presence of M-CSF, whereas no stimulating effect could be found when soluble RANKL was added to co-cultures (90). CD8[+] T cells were reported to increase the numbers of osteoclasts *in vitro* even in the presence of M-CSF and RANKL (90). The activating effect of T cells on osteoclast differentiation could also be boosted by cytokines like, IL-1, IL-18 and TNFα that up-regulate surface RANKL expression on T cells and thereby driving osteoclast differentiation (91). Until recently, the enhancing or suppressing effects of T cells has only been analyzed by using the entire $CD4^+$ or even $CD3^+$ T cell population, not taking into account different T helper (Th) cell subsets. Sato and colleagues recently published that both Th1 and Th2 cells had an inhibitory effect on osteoclast differentiation when co-cultured in the presence of M-CSF and RANKL (92). This inhibitory effect was based on the cytokines produced by these Th subsets such as IFNγ and IL-4. But even more interesting was the observation that Th17 cells enhance osteoclast differentiation by secretion of IL-17. The effect of IL-17 was not a direct effect on osteoclastogenesis but rather an indirect one through upregulation of RANKL on osteoblasts and bone marrow stromal cells (92). This observation is consistent with a previous report that IL-17 drives osteoclast differentiation by the induction of RANKL expression on osteoblasts (93). Not only T cells were reported to act on bone cells but also B cells. Also activated $CD19^+$ B cells can induce phenotypically smaller but highly active osteoclasts *in vitro* (90). These activated B cells expressed RANKL and TNFα, both known as inducers of osteoclastogenesis. Furthermore, B cells also express MIP-1α and MCP-3 (90), among which MIP-1α has been shown to induce migration of osteoclasts and together with MCP-3 stimulate osteoclast differentiation (94, 95). In contrast, did the group of Weitzman and Pacifici claim that B cells have suppressive effects on osteoclast differentiation due to their high production of OPG. This finding is supported by B cell deficient (μMT/μMT) mice which show an osteoporotic bone phenotype that could be rescued after B-cell reconstitution due to their high expression levels of OPG (96).

So far, most studies pointed towards a supportive action of $CD3^+$ T cells on osteoclast differentiation. These observations were commonly based on the expression of RANKL and TNFα by activated T cells. But when $CD3^+$ T cells were separated into the T helper cell subpopulations it was found that only Th17 cells and not Th1 or Th2 are the osteoclastogenic helper T cell subset that drives osteoclast differentiation. Moreover, another $CD3^+$ T cell population, the naturally occurring Treg and their role in bone homeostasis have not been characterized. However, recently we and others could show that addition of Tregs to monocyte cultures *in vitro* inhibits their differentiation into osteoclasts indicating that Tregs could suppress bone resorption (97-99). In contrast to a former study (92), all three studies describe the suppressive potential. But they show discrepancies in the mechanism of suppression by Tregs on osteoclastogenesis. In the present thesis we try to highlight the importance of direct cell-contact between Tregs and osteoclasts *in vitro* and *in vivo* as the main regulatory mechanism without disregarding the additional suppressive effects of cytokines secreted by Tregs.

2. Materials and Methods

2.1. Materials

RBC lysis buffer:
Tris (Merck, Cat.No. 1.08387.2500)
NH4Cl pro analyses (Merck, Cat.No. 1145.1000)

Kits:
Fixation and permeabilization kit buffers:
Fixation/Permeabilization Concentrate (eBioscience, Cat.No. 00-5123)
Fixation/Permeabilization Diluent (eBioscience, Cat.No. 00-5223)
Permeabilization Buffer (10x) (eBioscience, Cat.No. 00-8333)

$CD4^+CD25^+$ Regulatory T Cell Isolation Kit, mouse (Miltenyi Biotec, Cat. No. 130-091-041):
Cocktail of biotin-conjugated monoclonal anti-mouse antibodies against:
CD8a (Ly-2; isotype: rat IgG2a), CD11b (Mac-1; isotype: rat IgG2b), CD45R (B220; isotype: rat IgG2a), CD49b (DX5; isotype: rat IgM), Ter-119 (isotype: rat IgG2b). MicroBeads conjugated to monoclonal anti-biotin antibody (isotype: mouse IgG1) Monoclonal anti-mouse CD25 antibody conjugated to R-Phycoerythrin (PE) (clone: 7D4; isotype: rat IgM), MicroBeads conjugated to monoclonal anti-PE antibodies (isotype: mouse IgG1)
CD11b MicroBeads, human and mouse (Miltenyi Biotec, Cat. No. 130-049-601)

Mouse Th1/Th2 10plex FlowCytomix Multiplex Kit (Bender Medsystems, Cat.No. MS820FF) for:
GM-CSF, IFN-gamma, IL-1alpha, IL-2, IL-4, IL-5, IL-6, IL-10, IL-17, TNF-alpha
Acid Phosphatase, Leukocyte (TRAP) Kit (Sigma-Aldrich, Cat.No. 387A):
Acetate Solution (Sigma-Aldrich, Cat.No. 386-3)
Citrate Solution (Sigma-Aldrich, Cat.No. 91-5)
Fast Garnet GBC Base Solution (Sigma-Aldrich, Cat.No. 387-2)
Naphthol AS-BI Phosphoric Acid (Sigma-Aldrich, Cat.No. 387-1)
Sodium Nitrite Solution (Sigma-Aldrich, Cat.No. 91-4)
Tartrate Solution (Sigma-Aldrich, Cat.No. 387-3)

Media

Complete media:

alphaMEM+Glutamax (Gibco, Cat.No. 32561-029)

10% heat-inactivated FCS (Gibco)

1% penicillin/streptomycin antibiotics (Gibco, Cat.No. 15410-114)

Bone marrow transplantation media:

Medium 199 (Sigma, Cat.No. M4530)

Hepes buffer (Gibco, Cat.No. 15630-056)

DNAse (Sigma, Cat.No. D4527)

Gentamycin (Sigma, Cat.No. G1397)

Cytokines

Recombinant murine M-CSF (R&D Systems Inc., Cat.No. 416-ML)

Recombinant murine RANKL (R&D Systems Inc., Cat.No. 462-TEC)

Cell culture plates:

Corning® HTS Transwell® 96 well permeable support (Sigma-Aldrich, Cat.No. CLS3391)

BioCoat™ Osteologic™ (BD Bioscience, Cat.No. 354609)

FACS

CD3e-PerCP (BD Pharmingen, Cat.No. 553067)

CD4-PE-Cy7 (BD Pharmingen, Cat.No. 552775)

CD11b-Fitc (BD Pharmingen)

CD19-PE (BD Pharmingen)

CD25-APC (BD Pharmingen, Cat.No. 558643)

IL17A-Alexa Fluor 488 (BD Pharmingen, Cat.No. 560220)

Foxp3- Alexa Fluor 488 (Clone FJK-16s, eBioscience, Cat.No. 53-5773)

Annexin V (BD Pharmingen)

Anti-BrdU (BD Pharmingen)

Alexa Fluor 488 Rat IgG1,$_k$ Isotype Control (BD Pharmingen, Cat.No. 557856)

Cell-culture antibodies

IL-4R antibody (Clone mIL4R-M1, BD Biosciences)

IL-10R antibody (Clone 1B1.3a, BD Biosciences, Cat.No. 550012)

TGF-βRII antibody (synthetic peptide specific for p75 TGF-beta Type II receptors, Abcam, Cat.No. ab32798)

CTLA-4 (Clone UC10-4B9, eBioscience)

CD80 (16-10A1, eBioscience, Cat.No. 16-0801)

CD86 (PO3.1, eBioscience, Cat.No. 16-0861)

CTLA4-Ig (Bristol-Myers Squibb)

RT-PCR

DNAse I (Fermentas, Cat.No. EN0521)

TRIzol™ (Invitrogen, Cat.No. 15596-018)

RNAse inhibitor (Applied Biosystems, Cat.No. MP3247)

RT-PCR primer sequences:

foxp3 (sense 5'-AGG AGC CGC AAG CTA AAA GC-3', antisense 5'-TGC CTT CGT GCC CAC TGT-3')

RANKL (sense 5'-GAA TCC TGA GAC TCC ATG AAA ACG-3', antisense 5'-CCA TGA GCC TTC CAT CAT AGC TGG-3')

MMP9 (sense 5'-CAT TCG CGT GGATAA GGA-3', antisense 5'-TCA CAC GCC AGA AGA ATT TG-3')

TRAP (sense 5'-CGA CCA TTG TTA GCC ACA TAC G -3', antisense 5'-TCG TCC TGA AGA TAC TGC AGG TT-3')

NFATc1 (sense 5'-CCC GTT GCT TCC AGA AAA TA-3', antisense 5'-TCA CCC TGG TGTV TCT TCC TC-3')

Oscar (sense 5'-TCG CTG ATA CTC CAG CTG TC-3', antisense 5'-ATC CCA GGA GTC ACA ACT GC-3')

Cathepsin K (sense 5'-ATA TGT GGG CCA GGA TGA AAG TT-3', antisense 5'-TCG TTC CCC ACA GGA ATC TCT-3')

OPG (sense 5'-AGC TGC TGA AGC TGT GGA A -3', antisense 5'-GGT TCG AGT GGC CGA GAT-3')

β-actin (sense 5'-TGT CCA CCT TCC AGC AGA TGT-3', antisense 5'AGC TCA GTA ACA GTC CGC CTA GA-3')

hPRT (sense 5'-GTT AAG CAG TAC AGC CCC AAA-3'; antisense 5'-AGG GCA TAT CCA ACA ACA AAC TT-3')

RORγt (sense 5'-GCC TAC AAT GCC AAC AAC CAC ACA-3'; antisense 5'-ATT GAT GAG AAC CAG GGC CGT GTA-3')

Western Blot

Foxp3 antibody (Clone FJK-16s, eBioscience)

RANKL (R&D Systems, Cat.No. AF462

polyclonal rabbit anti-rat immunoglobulins (DakoCytomation, Denmark) indolamine 2,3-dioxygenase (IDO) antibody (Clone 10.1, Millipore, Cat.No. 05-840)

ELISA

mouse OPG (R&D Systems Inc)

mouse RANKL (R&D System Inc)

mouse Osteoclacin (Nordic Bioscience)

mouse C-terminal telopeptide α1 chain of type I collagen (CTX-I) (Ratlaps, Nordic Bioscience)

human TRAP5b (Quidel)

human Osteocalcin (Quidel)

human C-terminal telopeptide α1 chain of type I collagen (CTX-I) (CrossLaps, Nordic Bioscience)

Immunohistology

Foxp3 (eBioscience)

CD3 (Clone CD3-12, AbD Serotec, Cat.No. MCA1477)

CD45R (Clone RA3-6B2, Bd Pharmingen, Cat.No. 550286)

siRNA transfection

OptiMEM I Reduced Serum Medium (Invitrogen, Karlsruhe, Germany)

IDO specific siRNA (Stealth™ Select RNAi, Invitrogen, Karlsruhe, Germany)

negative control siRNA (Stealth™ RNAi Negative Control Duplexes, Invitrogen)

Lipofectamin RNAiMAX (Invitrogen, Karlsruhe, Germany)

Block-it Alexa Flour Oligo (Invitrogen, Karlsruhe, Germany)

2.2. Methods

Animals

The foxp3-transgenic (foxp3tg) mice (strain 2826; C57BL/6), the scurfy mutant (sf/Y), the human TNF transgenic (*hTNF*tg) mice (strain Tg197), $CD80/86^{-/-}$, $CD28^{-/-}$, $ICOS^{-/-}$ and $ICOSL^{-/-}$ were previously described (100-103). All animals were maintained in a SPF facility. All animal experiments were performed with the agreement of the ethic local authorities.

Isolation and characterization of regulatory T cells (Tregs)

Spleens were isolated from 5-8 week old female or male C57Bl/6 (wild type) mice and homogenized through a 70 µm stainless steel mash to archive a single-cell suspension. Erythrocytes in the cell suspension were lysed by resuspending the cell pellet with 5ml Red Cell Lyis Buffer (RBC) for 2-3 min at room temperature (RT). Lysis was stopped by adding 30ml PBS containing 10% heat inactivated FCS. $CD4^+CD25^-$ and $CD4^+CD25^+$ T cell populations were isolated from the spleenic cell suspensions using microbead-based Regulatory T Cell Isolation Kit (Miltenyi Biotec, Germany) according to the manufacture's instructions. The purity of the isolated cells was assessed by flow cytometry analyses (fluorescence- activated cell sorting (FACS)) of surface molecules (see below). Typically, $\geq$ 95% of the $CD25^+$ T cells eluted from LS column (Miltenyi Biotec, Germany) were positive for the intracellular staining of the transcription factor Foxp3. More than 98% of the cells were viable after purification.

Suppression assay of T cell proliferation

$CD4^+CD25^+$ and $CD4^+CD25^-$ T cells were stimulated either separately (5 x 10^4 cells/well) or in a co-culture at different ratios (5 x 10^4/ well $CD4^+CD25^+$ and 5 x 10^4/ well $CD4^+CD25^-$ per well (1:1 ratio)) with soluble anti-CD3e (5 µg/ml) (eBioscience) in the presence of irradiated (30 Gy) autologous feeder cells (1 x 10^5 cells/well) in 96-well round-bottom plates (Corning) in a total volume of 200 µl/well. Cells were cultured for 4 days and afterwards pulsed with 1 µCi of 3H-thymidine for 16 hours. Proliferation was determined by 3H-thymidine incorporation and was measured by a liquid scintillation counter (1205 Betaplate, Wallac Pharmacia). Each sample was performed in triplicate. The $CD4^+CD25^+$ T cell suppression capacity was assessed from the proliferation values of $CD4^+CD25^-$ T cells alone and in the co-culture with $CD4^+CD25^+$ T cells.

Flow cytometry (FACS analyses)
Analyses of the expression of cell surface molecules or of intracellular cytokines and proteins on a single cell level was performed by flow cytometry with a FACSCalibur (BD Biosciences) flow cytometer. Fluorescence signals were determined using appropriate electronic compensation to exclude emission spectra overlap.

FACS analyses of surface molecules
For extracellular staining, 1×10^6 cells/staining were washed with 1 ml FACS buffer, re-suspended in 100 µl FACS buffer and incubated with saturating amounts of PE-, FITC-, Alexa488-, PE-Cy5 or APC labelled antibody for 30 min at 4°C in the dark. One aliquot of the cells was parallel stained in parallel with corresponding control isotype antibody to monitor for the specificity of the staining. Afterwards, cells were washed twice with 1 ml FACS buffer, re-suspended in 300 µl FACS buffer and analyzed.

FACS analyses of intracellular molecules
Intercellular staining with 1×10^6 cells/staining was performed using the fixation and permeabilization kit for intracellular staining according to the manufacture's instructions (eBioscience). Annexin V and anti-BrdU were obtained from BD Biosciences.

Isolation and culture of osteoclast precursors
Bone marrow was isolated from 5-8 week old female or male C57Bl/6 (wild type), transgenic or knock-out mice line by flushing femoral bones with complete media. Erythrocytes in the cell suspension were lyzed by RBC buffer treatment as described before. Monocytes were isolated from bone marrow derived cell suspension using CD11b micro beads (Miltenyi Biotec, Germany) according to the manufactures instructions. The purity of isolated monocytes was assessed by flow cytometry analyses using CD11b-FITC labelled antibodies (Miltenyi Biotec, Germany). $CD11b^+$ monocytes were plated in 96-well plates (2.5×10^5/ well) or 48-well-plates (5×10^5/

well) in the presence of 30 ng/ml M-CSF and 50 ng/ml RANKL (R&D Systems Inc). Complete media was changed after 72 h. Osteoclast differentiation was evaluated by fixing and staining cells in the wells for tartrate resistant acid phosphatase (TRAP) using a Leukocyte Acid Phosphatase Kit (Sigma-Aldrich) according to the manufacture's instructions.

Co-culture of Tregs and osteoclasts
Purified $CD11b^+$ monocytes (2.5×10^5/ well) and different numbers of activated $CD4^+CD25^+$ Tregs ($5 \times 10^4 - 5 \times 10^3$/ well) or $CD4^+CD25^-$ T cells were co-cultured in 96-well plates in the presence of 30 ng/ml M-CSF and 50 ng/ml RANKL (R&D Systems Inc.) for 4-6 days. After 72h complete media was changed. T cells were activated with soluble anti-CD3e (5 µg/ml) (eBioscience) in the presence of irradiated (30 Gy) autologous feeder cells (1×10^5 cells/well) in 96-well round-bottom plates (Corning) in a total volume of 250 µl/well. In addition, to keep total numbers of cells constant in the co-culture experiments irradiated $CD4^-$ T cells were titrated in.

Transwell experiments
A similar approach as in the co-culture of Tregs and osteoclasts experiments was used. Briefly, 96-well plates (0.4µm pore size, Corning) were loaded with 2.5×10^5 $CD11b^+$ monocytes into the lower chambers and various numbers of $CD4^+CD25^+$ Tregs or $CD4^+CD25^-$ T cells into the upper chambers. T cells, in the upper chamber, were activated with soluble anti-CD3e (5 µg/ml) (eBioscience) in the presence of irradiated (30 Gy) autologous feeder cells (1×10^5 cells/well). Culture was performed in the presence of 30 ng/ml M-CSF and 50 ng/ml RANKL. Complete media was changed after 72 h.

Bone resorption experiments
Calcified matrix resorption activity of the osteoclasts was tested using BioCoat™ Osteologic™ plates (BD Biosciences) according to the manufactures instructions. Afterwards von Kossa stain was performed with the BioCoat™ Osteologic™ Discs to visualize the resorption pits. Resorption pits area was assessed using the Nikon NIS-Elements 3.0 Software.

Inhibition studies

For cytokine neutralization experiments blocking antibodies against IL-4R (Clone mIL4R-M1, BD Biosciences), IL-10R (Clone 1B1.3a, BD Biosciences) and TGF-βRII (synthetic peptide specific for p75 TGF-beta Type II receptors, Abcam) were used and were pre-incubated with CD11b$^+$ monocytes for 1h at 37°C before co-culture with CD4$^+$CD25$^+$ T cells. For surface molecule blocking experiments, anti CTLA-4 (Clone UC10-4B9, eBioscience), CD80 (16-10A1, eBioscience), CD86 (PO3.1, eBioscience) were pre-incubated with CD11b$^+$ monocytes or CD4$^+$CD25$^+$ T cells at different concentrations for 1h at 37°C. Osteoclast differentiation was evaluated by staining for tartrate resistant acid phosphatase (TRAP) using a Leukocyte Acid Phosphatase Kit (Sigma-Aldrich).

RT-PCR

Isolation of RNA was done with the standard combination of TRIzol^{TM} and phenol/chloroform procedure. RNA concentration was measured before genomic DNA was digested with DNAse I (Fermentas) and cDNA was synthesized in the presence of RNAse inhibitor. For SYBR Green-based detection, a dissociation curve was carried out by one cycle following the last amplification cycle to control for the specificity of PCR amplification: 95°C for 15 sec, 60°C for 30 sec, 95°C for 15 sec. The list of genes analyzed by RT-PCR and the SYBR Green primers are listed in Materials. Relative quantification was performed by calculating the difference in cross-threshold values (ΔCt) of the gene of interest and a housekeeping gene according to the formula $2^{-\Delta Ct}$. In some experiments, the relative expression values were normalized to the expression values in the control condition.

Western blot analyses

Spleens were isolated from C57Bl/6 (wild type) mice and single cell suspensions were generated. CD4$^+$ CD25$^-$ and CD4$^+$CD25$^+$ T cells populations were isolated from the splenic cell suspensions using microbead- based Regulatory T Cell Isolation Kit (Miltenyi Biotec, Germany) as describe above. T cell populations were lysed in Laemmli (BioRad, Germany)) sample buffer. Lysates representing 3 x 10^6 cells were separated on SDS-PAGE gels, transferred to nitrocellulose membrane and stained with anti-mouse Foxp3 antibody (Clone FJK-16s, eBioscience) following polyclonal rabbit anti-rat immunoglobulins (DakoCytomation, Denmark). Bone marrow derived

CD11b$^+$ monocytes were cultured at a concentration of 1x10^{6}cells/ml in the presence of 30 ng/ml M-CSF and 50 ng/ml RANKL (R&D Systems Inc.) and either challenged with different concentrations of CTLA4-Ig (Bristol-Myers Squibb) 5x10^5 CD4$^+$CD25$^-$ or CD4$^+$CD25$^+$ T cells. Cells were processed as described above and stained with anti-mouse indolamine, 2-3, dioxygenase (IDO) antibody (Millipore,CA) following polyclonal goat anti-rat immunoglobulins.

ELISA

CD4$^+$ CD25$^-$ and CD4$^+$CD25$^+$ T cells populations were activated in pre-coated anti-CD3e (eBioscience) monoclonal antibody (5 µg/ml) 96-well plates overnight. For detection of osteoprotegerin (OPG) cell culture supernatants were harvested after 24 h of culture and OPG content was measured by a quantitative sandwich ELISA for murine OPG (R&D Systems Inc). Serum measurements for murine RANKL (R&D Systems Inc), OPG (R&D Systems Inc), Osteocalcin and C-terminal telopeptide α1 chain of type I collagen (CTX-I) (Nordic Bioscience) were measured by ELISA. In addition, serum measurements for human TRAP5b (Quidel), Osteocalcin (Quidel) and C-terminal telopeptide α1 chain of type I collagen (CTX-I) (CrossLaps, Ratlaps, Nordic Bioscience) serum levels were also measured by ELISA. All ELISA analyses were performed according to the manufacturers' instructions.

µCT Imaging

µCT images of tibias were acquired on a laboratory cone-beam µCT scanner developed at the Institute of Medical Physics for ultra high resolution imaging (55). It uses a µ-Focus x-ray tube (Hamamtsu) and a 2D cooled CCD detector array (1024*1024 elements, 19µm pitch; Photometrics, USA) with a dynamic range of 16 bit. A fibre optics taper enlarges the sensitive input area of the CCD by a factor of three. Detector and sample stage can be linearly translated independently with respect to each other and with respect to the x-ray source providing variable magnification of the object. For the current project the following acquisition parameters were used: voltage: 40 kV, X-ray current: 250 µA, exposure time: 5000 ms/projection, 720 projections, matrix: 1024x1024, voxel size in reconstructed image: 9 µm. Images were analyzed using a plug-in programmed for Amira 4.1.2. (Mercury) with the histomorphometric parameters TV/BV, Tb.Th. and Tb.N were calculated. The tibia samples of CD80/86-/-, CD28-/-, ICOS-/- and ICOSL-/- were measured with a commercially available desktop MicroCT, (µCT35, SCANCO Medical AG, Brüttisellen, Switzerland) (56).

Bone histomorphometry

Histomorphometry was performed on methacrylate-embedded undecalcified plastic sections stained with von Kossa and Goldner for bone. Quantifications were performed by digital image analyses (OsteoMeasure, OsteoMetrics). Arthritis (paw swelling and grip strength) were assessed by a semi-quantitative score as described previously (57). Histological analyses were performed on formalin-fixed, decalcified, paraffin-embedded tissue sections stained with haematoxylin and eosin (H&E), tartrate-resistant acid phosphatase (TRAP) or toluidine blue. Synovial inflammation, osteoclast numbers and cartilage destruction were quantified by digital image analyses (OsteoMeasure, OsteoMetrics).

Immunohistology

Deparaffinised ethanol-dehydrated tissue sections were pre-treated with high temperature unmasking solution (20 min in citrate buffer, pH 6.0) for staining of Foxp3 (FJK-16s) (eBioscience), CD3 (BD Biosciences), CD45R (BD Biosciences), F4/80 (BD Biosciences) and Neutrophils (BD Biosciences).

Measurement of serum cytokines

Cytokines in serum/plasma were measured with the Mouse Th1/Th2 10plex FlowCytomix Multiplex Kit (Bender Medsystems) according to the manufacturer's instructions using a FACSCalibur (BD Biosciences) flow cytometer.

Bone marrow transplantation

Recipient *hTNF*tg mice (6 weeks old) were irradiated at 11 Gy using orthovoltage irradiation (Stabilipan, Siemens, Germany) at 250 kV / 15 mA / 40 cm focus – surface distance at a dose rate of 1.15 Gy/min. For irradiation the mice were anaesthetised by inhalation anaesthesia (Forene, Abbot, Germany). The anaesthesia was performed during the irradiation process in a closed fixture made of Plexiglas. This fixture was mounted on a Plexiglas-block (d = 50 mm) to achieve full reflexion scattering. The next day mice were reconstituted by intravenous injection of 5×10^6 bone marrow cells in Medium 199 (Sigma) containing 5ml 1M hepes buffer (Gibco),

5ml (1mg/ml) DNAse (Sigma) and 40µl (50mg/ml) gentamycin (Sigma), and analyzed 6 weeks after transplantation.

Ovariectomy

Mice were anesthetized with 200µl of anaesthesia (1ml Ketavet 100mg/ml, 0,5ml Rompun 2%, 8,5ml NaCl 0.9%) and either sham-operated or ovariectomized (ovx) at 6 weeks of age and analyzed at 12 weeks of age. One tibia was excised for histological analyses and one for µCT imaging. Taken ovaries were histologically identified to verify successful ovx.

Tryptophan / kynurenine measurement

Tryptophan and kynurenine were determined simultaneously using liquid chromatography tandem mass spectrometry, with atmospheric pressure chemical ionization in the positive ion mode (API 4000 Q Trap, Applied Biosystems, MDS Sciex). 10 µl of the samples and calibrators were deproteinized with sulfosalicylic acid (10%). LC was performed using a Chromolith column (RP-18e, 100 * 3.0 mm, Merck, Darmstadt, Germany) at a flow rate of 1ml/min with 2mM ammoniumacetate / methanol (78:22, v/v, pH = 2.0). The total run time was 2.5 min. Sample analyses was performed in the multiple-reaction monitoring mode with a dwell time of 100 ms per channel using the following transitions for quantification (qualifier transition): m/z 205.2/187.8 (205.2/146.0) tryptophan m/z 209.2/191.7 (209.2/146.0) kynurenine, m/z 210.2/192.8 tryptophan-d5 (internal standard).

Transfection with siRNA

Macs sorted, CD11b$^+$ monocytes were transfected in OptiMEM I Reduced Serum Medium (Invitrogen, Karlsruhe, Germany) for 6h with an IDO specific siRNA (Stealth™ Select RNAi, Invitrogen, Karlsruhe, Germany) or a negative control siRNA (Stealth™ RNAi Negative Control Duplexes, Invitrogen) using Lipofectamin RNAiMAX (Invitrogen, Karlsruhe, Germany) according to the manufacturer's instructions. Transfection efficiency was assessed by Block-it Alexa Flour Oligo (Invitrogen, Karlsruhe, Germany) positive control.

Blood donors

Serum levels of collagen type I cleavage products (CTX-I), tartrate resistant acid phosphatase (TRAP) 5b and osteocalcin were analyzed by ELISA (TecoMedical, Sissach, CH; for CTX-I: Nordic Biosciences, Herlev, DK) in 35 healthy subjects (mean±SEM age 55.8 ±8.1 years, 57% females) and 35 patients with rheumatoid arthritis (mean±SEM age 57.9±6.4 years, 63% females). Tregs were analyzed by Treg detection kit (Miltenyi Biotec, Bergisch-Gladbach, Germany) using antibodies against CD4, CD25 and Foxp3. Cells were gated for CD4 positive cells and Treg were identified as CD25high, Foxp3-positive cells. Studies were approved by the local ethical committee of the University of Erlangen and written informed consent was given by all blood donors.

Statistical analyses

All statistical analyses were performed with Student's t-test, one-way or two-way ANOVA followed by Tukey`s test and are represented as means ± s.e.m. unless otherwise stated using GraphPad Prism 4.0 Software. * indicates $P < 0.05$; ** indicates $P < 0.01$, *** indicates $P < 0.001$. All experiments were done with n=10 mice per group unless otherwise stated.

3. Results

3.1. Suppressive effect of Tregs on osteoclastogenesis *in vitro*

3.1.1. Isolation and characterization of Tregs and osteoclast precursors

For co-culture experiments of Tregs with osteoclasts we isolated $CD4^+CD25^+$ T cells with magnetic bead separation from the spleen of healthy C57Bl/6 mice. Purity of $CD4^+CD25^+$ T cells after isolation was more than 90% **(Figure 1B)**. The transcription factor Foxp3 is expressed in most of the human and in all murine Tregs and is therefore considered as a specific Tregs marker. We thus assessed the expression of Foxp3 in permeabilized cells to evaluate the proportion of Tregs present in the pool of isolated CD4+CD25+ T cells. Virtually all of the cells (> 93%) also expressed Foxp3 **(Figure 1A)**. We also analyzed the expression of *foxp3* mRNA, and Foxp3 protein showing that $CD4^+CD25^+$ T cells expressed significantly higher amounts of Foxp3 **(Figure 1D and E)**. Next we analyzed the functionality of the isolated $CD4^+CD25^+Foxp3^+$ T cells (Tregs) and showed that these cells efficiently suppress proliferation of $CD4^+CD25^-$ T cells in standard suppression assay experiments (i.e. T cells were stimulated by anti CD3e soluble antibody and irradiated $CD4^-$ cells as antigen presenting cells (APC)). The observed effect was dose dependent, indicating that the isolated Tregs were functionally competent Tregs **(Figure 1C)**.

Osteoclast precursors were isolated with magnetic bead separation for CD11b from bone marrow cells of the same healthy C57Bl/6 mice. Purity of sorted cells was assessed by flow cytometry analyses in FSC and the expression of CD11b, which has been described as marker for the monocyte population capable to differentiate into osteoclasts after stimulation with M-CSF and RANKL (Ping L., A&R, 2004) **(Figure 2A)**. In addition, osteoclast differentiation and bone resorption assays with M-CSF alone, as control, or with M-CSF and RANKL was performed to show the differentiation potential of $CD11b^+$ bone marrow cells **(Figure 2B and C)**.

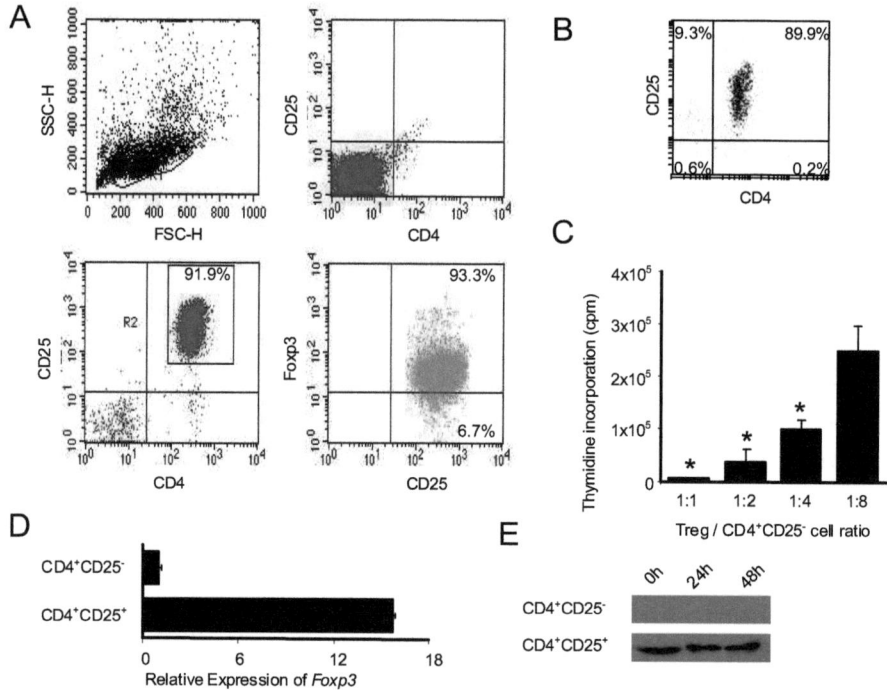

Figure 1. Isolation, purification and characterization of regulatory T cells

(A) Spleen cells from C57Bl/6 mice were sorted for CD4 and CD25 in DakoCytomation MoFlow and stained for CD4,CD25 followed by anti-Foxp3 intracellular staining and analyzed by flow cytometrie (FACS). (B) FACS analysis of CD25$^+$CD4$^+$ (Tregs) T cell population isolated from the spleenic cell suspensions using microbead based Tregs Isolation Kit for expression of CD4 and CD25. Data are presented as a representative dot plot from one experiment. (C) Analysis of suppressor activity of Tregs isolated from WT mice in co-cultures with freshly isolated CD4$^+$CD25$^-$ responder T cells in the presence of anti-CD3 and irradiated APC. Proliferation measured by [3H]thymidine incorporation during the last 18h of a 3-day culture period. Representative results of one of three independent experiments are shown. (D) Relative expression of *Foxp3* mRNA in Tregs and CD4$^+$CD25$^-$ T cells purified from spleen. (E) Western blot analysis of Foxp3 expression in purified WT CD4$^+$CD25$^-$ T cells and Tregs either freshly isolated (0h) or cultured for 24 h and 48 h in the presence of plate bound anti-CD3.

3.1.2. Regulatory T cells suppress osteoclast formation

We next addressed whether regulatory CD4$^+$CD25$^+$Foxp3$^+$ T cells (Tregs) could suppress osteoclast formation. As already shown, in absence of T cells, osteoclast formation was induced by cultivating CD11b$^+$ osteoclast precursors with MCSF and

RANKL leading to the formation of multinucleated TRAP⁺ osteoclasts within 4-6 days of cell culture **(Figure 2)**.

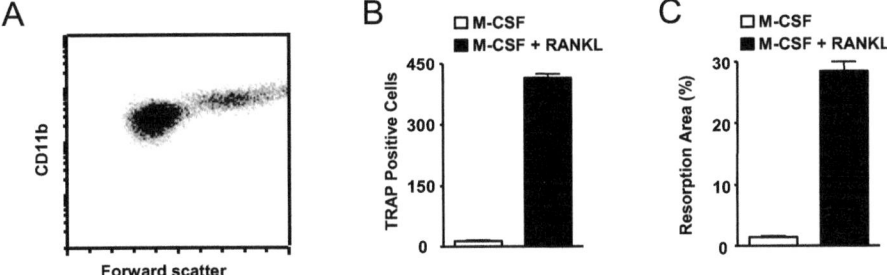

Figure 2. Isolation, differentiation and functional characterization of osteoclasts

(A) Flow cytometry (FACS) analysis of purified bone marrow derived monocytes from wild-type (WT) mice using microbeads for CD11b surface markers for forward-scatter and expression of CD11b⁺. **(B)** Osteoclast differentiation assay from WT purified bone marrow derived CD11b⁺ monocytes analysed for multinucleated TRAP positive stained cells. **(C)** Bone resorption assay on Osteologic™ plates from WT purified bone marrow derived CD11b⁺ monocytes analysed for resorpted bone matrix.

Addition of Tregs to unsorted, complete bone marrow cells with M-CSF and RANKL at day 1 dose-dependently suppressed osteoclast formation **(Figure 3A)**. To identify the responsible suppressive T cell subpopulation, sorted splenic CD4⁺ T cells were compared to Tregs. Inhibition of osteoclast formation from CD11b⁺ bone marrow cells was only observed in the presence of a high proportion of CD4⁺ CD25⁺ Tregs (1 per 5 osteoclast precursors), which reflects the 5% fraction of Tregs present within the CD4⁺ T cell pool **(Figure 3B)**. We next asked whether Tregs need to be activated in order to suppress osteoclast differentiation. In contrast to activated Tregs, non-activated Tregs failed to inhibit osteoclast formation except when they were added in high concentrations **(Figure 3C)**.

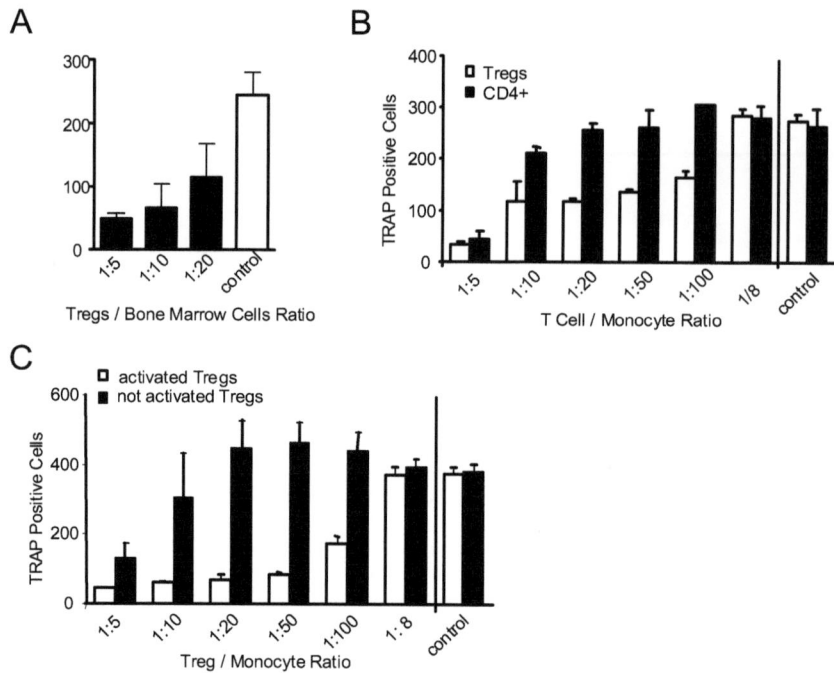

Figure 3. Activated CD4⁺CD25⁺ T cells (Tregs) are the responsible T cell population for the suppressive effect on osteoclast differentiation.

(A) Tregs suppress osteoclast differentiation from complete bone marrow cells. (B) Suppression is enhanced with the Tregs population compared to the whole CD4⁺ T cell population. Purified monocytes were either co-cultured with activated or purified Tregs or CD4⁺ T cells at identical absolute cell numbers in the presence of M-CSF/RANKL. (C) Suppression is more efficient with activated Tregs than with non-activated Tregs. Purified monocytes were either co-cultured with activated or not activated purified Tregs in the presence of M-CSF/RANKL. Control shows monocytes with CD4⁺CD25⁻ T cells at a ratio of 1:5.

There was a 20-fold difference in the inhibitory potential between activated and non-activated Tregs. Even small numbers of Tregs were sufficient to significantly blunt osteoclast formation and Tregs could virtually abolish the formation of osteoclasts **(Figure 4A)**. In contrast, CD4⁺CD25⁻ T cells did not affect the osteoclast formation as shown by the control bars **(Figure 4A)**. Representative images of a Treg-monocyte co-culture experiment is shown in **Figure 4B**: Differentiated osteoclasts which specifically expressed high amount of TRAP appeared as multinucleated purple giant cells, whereas osteoclast precursors are mononuclear TRAP positive cells. Addition of Tregs dramatically inhibits the formation of TRAP positive multinucleated cells in a dose dependent fashion. Suppression of osteoclast differentiation in this co-culture system was effective when Tregs were present along the entire culture period or added during the 3 first days. When added at day 4 or day 5, no suppressive effect of Tregs on osteoclasts was observed **(Figure 4C)**. In order to confirm that the effects of Tregs addition were due to specific inhibition of osteoclast formation and not due to simply crowding out osteoclast precursors, the total numbers of cells was kept constant by complementing the culture with $CD4^+CD25^-$ T cells **(Figure 4D)**. We then investigated the functional consequences of impaired osteoclast formation on their activity i.e. bone resorption. Therefore, we performed the same experimental setting as described for the analyses of $TRAP^+$ cell counts. This time, the cells were plated on Osteologic^{TM} Bone Cell Culture System wells, which represent cover-slips coated with hydroxyapatite (calcium phosphate), the main mineral component of bone. This system is a convenient alternative to cortical or ebony chips commonly used for direct assessment of osteoclast activity *in vitro*. Whereas resorption pits were found upon differentiation of osteoclasts, such lesions were absent when Tregs were added **(Figure 4E and F)**. This suggests that $CD4^+CD25^+foxp3^+$ Tregs not only suppress osteoclast formation but also functional bone resorption since bone resorption decreases with decreased numbers of mature bone resorbing osteoclasts.

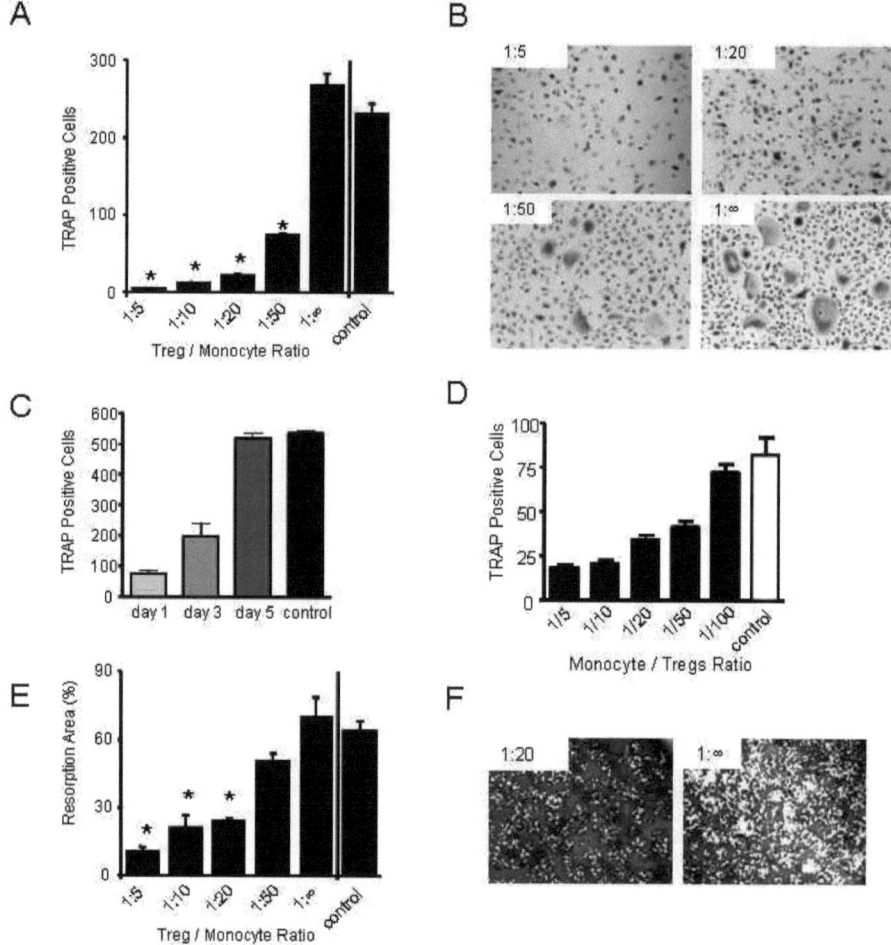

Figure 4. Regulatory T cells inhibit osteoclast differentiation and bone resorption

(A) CD4$^+$CD25$^+$Foxp3$^+$ T cells (Tregs) mediate the regulation of osteoclastogenesis from sorted CD11b$^+$ monocytes in a dose dependent manner. Control shows monocytes co-cultured with CD4$^+$CD25$^-$ T cells at a ratio of 1:5. (B) Dose dependent suppression of osteoclastogenesis by Tregs. Representative microphotographs of TRAP positive stained *in vitro* co-cultures shown in (A); original magnification 20x. (C) Time dependent suppression of osteoclast differentiation in co-culture experiments with Tregs. (D) Co-culture of CD11b$^+$ WT monocytes with activated WT Tregs. In this experiment the number of T cells per well was kept constant by filling up the decreasing number of Tregs with CD4$^+$CD25$^-$ T cells. (E) Tregs control the bone resorbing activity of osteoclasts. Analysis of resorption activity measured on Biocoat™ Osteologic™ Disc with similar experimental settings as in (B). (F) Representative pictures of van Kossa stained Osteologic™ discs at different Treg /monocyte ratios. White dots display resorption area.

Profiling of mRNA expression of osteoclast-differentiation associated markers showed reduced expression of *Trap, Oscar, Cathepsin K, Mmp-9,* and *Nfatc1* in CD11b$^+$ osteoclast precursor cells co-cultured with Tregs as compared to co-cultures with CD4$^-$CD25$^-$ T cells **(Figure 5)**. This confirmed the direct suppression by activated Tregs on differentiation of osteoclast precursors to mature bone resorbing multinucleated osteoclasts.

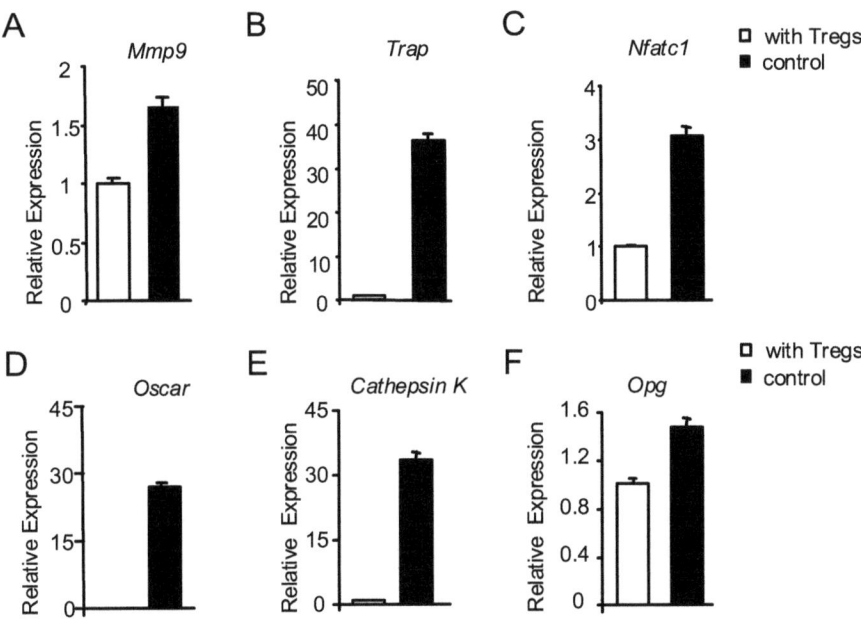

Figure 5. Characterization of counted TRAP positive osteoclasts

Relative expression of osteoclast-differentiation markers in monocytes / osteoclasts. Bone marrow derived monocytes were cultured in the presence of M-CSF and RANKL either with or without (control) CD4$^+$CD25$^+$ T cells (Tregs) activated with anti-CD3 and irradiated APC. After 4 days mRNA of monocytes/osteoclasts was isolated and analyses by quantitative PCR. Relative gene expression was derived from the ratio of gene of interest to β-actin mRNA expression. **(A)** *Mmp9,* **(B)** *Trap,* **(C)** *Nfatc1,* **(D)** *Oscar,* **(E)** *Cathepsin K* and **(F)** *Opg* mRNA.

3.1.3. Direct cell-contact is essential for suppression

We next questioned the mechanisms of suppression on osteoclastogenesis by activated Tregs. We first investigated if activated Tregs were able to modify the ratio in levels of RANKL to its decoy receptor OPG. Interestingly, while Tregs as well as $CD4^+CD25^-$ T cells expressed mRNA of RANKL, its expression was even increased in activated Tregs compared to $CD4^+CD25^-$ T cells **(Figure 6A)** suggesting a potential positive effect of Tregs on osteoclastogenesis. However, an increase in OPG expression was also measured that most likely compensated for the increased RANKL level **(Figure 6A)**. In addition, there was no increase in the production of OPG protein in the supernatants of the Tregs / monocytes co-cultures. Thus a change in the ratio RANKL/OPG could not explain the suppression of osteoclastogenesis by activated Tregs **(Figure 6B)**.

Numerous cytokines produced by T cells such as IL-4 and IL-10 have been shown to inhibit osteoclastogenesis *in vitro*. Therefore, we performed co-cultures of Tregs with monocytes and tested the rescue potential of neutralizing antibodies against TGFβ, IL-4 and IL-10 on the suppressive effect on osteoclast differentiation by Tregs. The addition of any of these antibodies alone or in combination rescued osteoclast differentiation to a significant level ($P < 0.05$) but no full rescue could be achieved **(Figure 6C)**.

To determine whether direct cell-contact was necessary for the suppressive effect of Tregs on osteoclastogenesis, co-cultures of Tregs with monocytes were performed in a Transwell cell culture systems preventing direct cell contact between Treg and osteoclast precursors. $CD11b^+$ sorted monocytes were seeded in the lower chamber of Transwell plates and Tregs together with a constant number of irradiated APCs were titrated at different ratios into the upper chamber of the Transwell plates. The suppressive effect of Tregs on osteoclast formation was completely abolished when cells were not allowed to directly interact to each other **(Figure 6D)**. This indicates that direct cell-contact is an essential prerequisite for Tregs to inhibit osteoclast differentiation from $CD11b^+$ monocyte precursor cells.

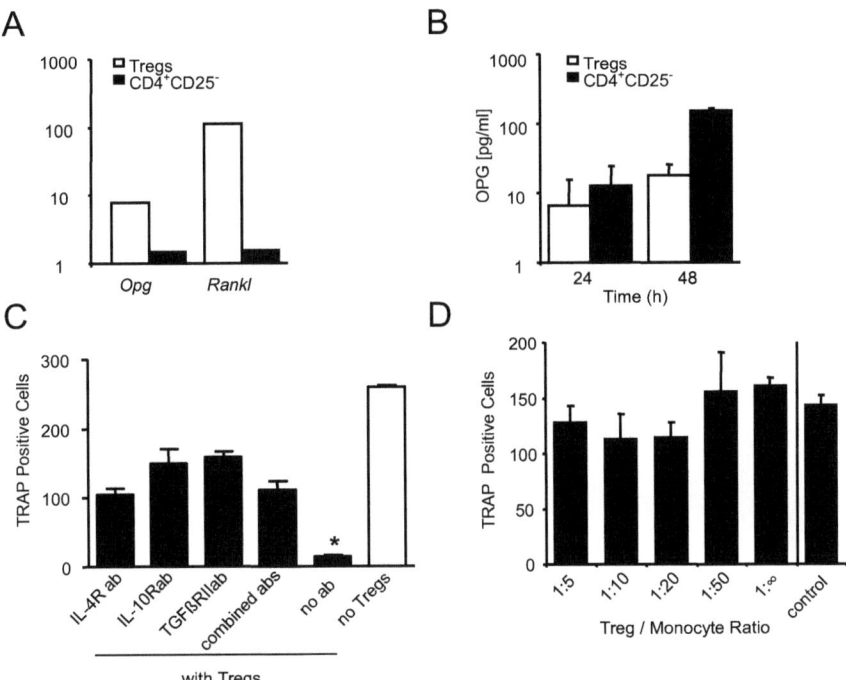

Figure 6. Inhibition of osteoclast differentiation by CD4⁺CD25⁺ T cells (Tregs) is cell contact dependent

(A) Relative expression of *Opg* and *Rankl* mRNA in Tregs and CD4⁺CD25⁻ T cells purified from WT spleens were analyzed by RT-PCR. (B) OPG ELISA analysis in the supernatants of cultured Tregs or CD4⁺CD25⁻ T cells after 24h and 48h of activation in the presence of anti-CD3 monoclonal antibodies. (C) Specific blocking of cytokine receptors partially rescues osteoclast differentiation. For neutralization experiments blocking antibodies against IL-4R, IL-10R and TGF-βRII were used, either alone or all three together, and added to purified monocytes 1h before co-culture with Tregs. Mean values (± SD) from three different experiments were shown. (D) Inhibitory effect of Tregs on osteoclast differentiation in 96-well Transwell plates. Bone marrow derived monocytes (lower chamber) stimulated with RANKL/M-CSF were co-cultured with activated spleenic derived Tregs (upper chamber) at different ratios. Control shows monocytes with CD4⁺CD25⁻ T cells at a ratio of 1:5. Mean values (± SD) from triplicate co-cultures were shown.

3.1.4. CTLA-4 interaction with CD80/86 is mediating the suppression of osteoclast differentiation by Tregs

Based on the pivotal role of cell contact in the suppressive effects of Tregs on osteoclast differentiation, we hypothesized that CTLA-4 (CD152), that is constantly expressed by Tregs might be involved in this process. The addition of CTLA-4 Ig that mimic CTLA-4 activity to osteoclast precursor cells dose dependently suppressed osteoclast differentiation **(Figure 7A)**. This process could be completely reversed by adding anti–CTLA-4 antibodies, regardless of whether CTLA-4 Ig **(Figure 7A and B)** or Tregs **(Figure 8A)** were used to suppress osteoclast differentiation, indicating that the effect on osteoclast precursors is mainly mediated by CTLA-4.

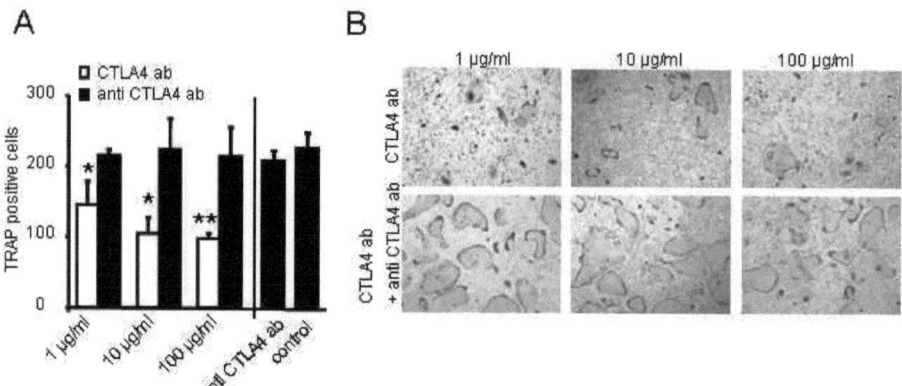

(A) Dose dependent suppression of osteoclast formation by CTLA-4 Ig. Different concentrations of CTLA-4 Ig or anti CTLA-4ab pre-incubated with CTLA-4 Ig were added to purified bone marrow derived monocytes stimulated with M-CSF/RANKL. Control shows monocyte culture without any treatment or just with the anti CTLA-4ab. **(B)** Representative microphotographs of *in vitro* cultures (TRAP staining) from Figure 7A; original magnification 20x.

CTLA-4 is a ligand for the T cell co-stimulatory molecules CD80/86, a pair of co-stimulatory molecules expressed on the surface of monocytes. Blockade of CD80/86, by neutralizing antibody fully abolished the inhibitory effect of Tregs on osteoclastogenesis **(Figure 8B)**. Moreover, osteoclast precursors from CD80/CD86 double knockout mice (*CD80/86$^{-/-}$*) were fully resistant to the inhibitory effects of Tregs, suggesting that CD80/86 is required for the regulation of bone resorption by Tregs **(Figure 8C and D)**.

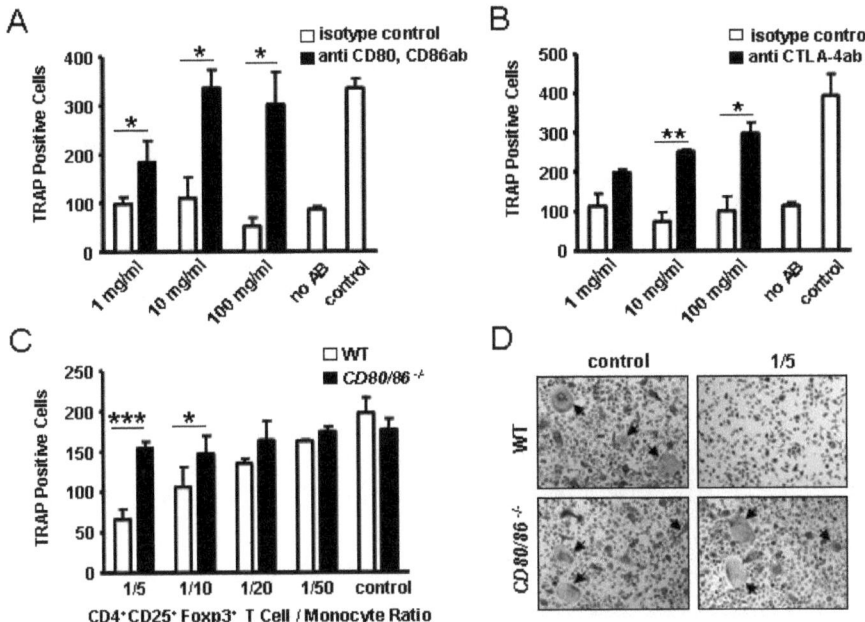

Figure 8. CTLA-4 and CD80/86 are essential for Foxp3 positive Treg suppression of osteoclastogenesis.

Co-culture of CD4$^+$CD25$^+$Foxp3$^+$ T cells from WT mice with both **(A)** anti CD80 and anti CD86 mAb and **(B)** anti-CTLA-4 mAb or isotype controls. **(C-D)** Osteoclast differentiation of CD11b$^+$ monocytes from WT and *CD80/86$^{-/-}$* mice co-cultured with activated CD4$^+$CD25$^+$Foxp3$^+$ T cells from WT mice. Osteoclasts are indicated by arrows.

This effect was specific to CD80/86, as the absence of another molecule expressed on osteoclast precursors and involved in T cell co-stimulation, ICOSL, did not affect Treg-mediated inhibition of osteoclastogenesis **(Figure 9A)**. To determine if this process is affected by CD28 or ICOS, Tregs were isolated from *CD28$^{-/-}$* and *ICOS$^{-/-}$* mice and these Tregs did effectively suppress osteoclast formation **(Figure 9B and C)**. In summary, we could show that Tregs are potent suppressors of osteoclast differentiation *in vitro*. Tregs suppress early differentiation steps, since at the late stage of differentiation or on mature osteoclasts they did not exert their suppressive capacity. Direct cell-contact of CTLA-4 on Tregs with CD80/86 on osteoclast precursors is essential for suppression of osteoclast differentiation. The cytokines expressed by Tregs, such as IL-4 or IL-10 and TGFβ show an additional suppressive effect but their effect is still dependent on cell to cell contact.

Results

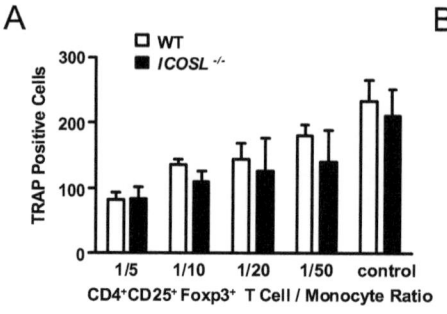
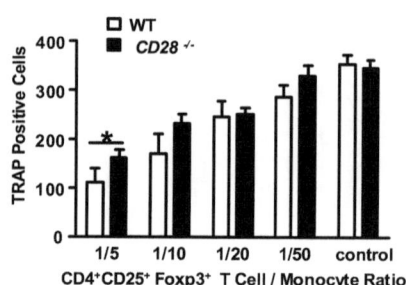
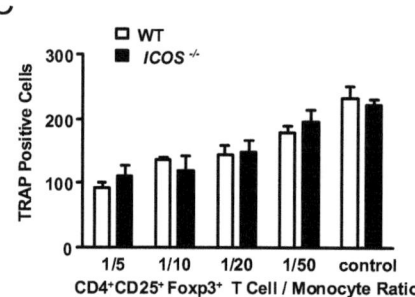

Figure 9. ICOSL, CD28 and ICOS are not essential for the suppressive effect of Tregs on osteoclastogenesis.

(A) Osteoclast differentiation assay *in vitro* of CD11b$^+$ monocytes from *ICOSL* $^{-/-}$ mice co-cultured with activated CD4$^+$CD25$^+$Foxp3$^+$ T cells (Tregs) from WT mice. Osteoclast differentiation assay *in vitro* of CD11b$^+$ monocytes from WT mice co-cultured with activated Tregs from **(B)** *CD28* $^{-/-}$ mice or **(C)** *ICOS* $^{-/-}$ mice.

3.2. The role of Tregs in bone homeostasis

3.2.1. Increased bone mass in *foxp3*-transgenic (*foxp3*tg) mice

Having shown that Tregs can suppress osteoclast formation *in vitro*, we analyzed their role in bone homeostasis *in vivo*. We first investigated whether mice over-expressing the central transcription factor of Tregs, Foxp3, (*foxp3*tg) show an overt bone phenotype. *Foxp3*tg mice have increased numbers of CD$^+$CD25$^+$Foxp3$^+$ T cells in the spleen (104) as well as in the bone marrow as shown by flow cytometry and quantitative PCR analyses **(Figure 10A and B)**. Similar total splenic cell numbers and amounts of CD3$^+$ T cells were counted in *foxp3*tg mice compared to littermate controls (WT) **(Figure 10C and D)**, indicating that the increased Treg numbers in *foxp3*tg mice was not due to an overall increase in splenic cell numbers but is due to a higher proportion of CD4$^+$CD25$^+$foxp3$^+$ T cells within the CD3$^+$ T cell population. Histochemical cell counts of Foxp3$^+$ cells in the tibia of *foxp3*tg mice showed a significant increase in Foxp3$^+$ cells in the bone shaft and an even more pronounced increase in the epiphysis that is the site of intense bone remodeling compared to WT mice **(Figure 10E and F)**.

Micro-computed tomography (µCT) was performed to determine the effect of increased Tregs numbers on bone mass. Tibias from WT and *foxp3*tg sex-matched littermates were compared showing a significantly increased bone mass (BV/TV: bone volume per total volume) in *foxp3*tg mice **(Figure 11A and B)**. The phenotype was based on higher number of bony trabeculae **(Figure 11B)**. Decreased osteoclast numbers and osteoclast-covered bone surface indicated impaired bone resorption in *foxp3*tg mice **(Figure 11C)**, which was confirmed by reduced collagen type I cleavage (CTX-I) products in the serum **(Figure 11D)** and *Trap* mRNA expression in bones of *foxp3*tg mice **(Figure 11E)**. In contrast, bone formation parameters such as mineral apposition rate (MAR) **(Figure 11F)**, osteoblast numbers, osteoblast-covered bone surface and osteoid volume were unchanged in *foxp3*tg mice **(Figure 11G)**. Low osteoclast numbers in *foxp3*tg mice were not due to a decreased number of osteoclast precursors in the bone marrow as shown by similar amount of CD11b$^+$ monocytes present in bone marrow of WT and *foxp3*tg mice **(Figure 11H)**. It was also not due to an (i) intrinsic impaired ability of monocytes to differentiate into osteoclasts since normal osteoclast assays did not show any difference in *foxp3*tg mice **(Figure 12A)**, (ii) neither to an altered cytokine milieu in serum levels of the major osteoclastogenesis modulation cytokines RANKL and OPG **(Figure 12C)**, (iii) or to an altered Th1, Th2 serum cytokine profile **(Figure 12D,-H)**, (iv) nor to a changed Th17 cells proportion as indicated by similar mRNA expression levels of *RORγT* **(Figure 12I)**.

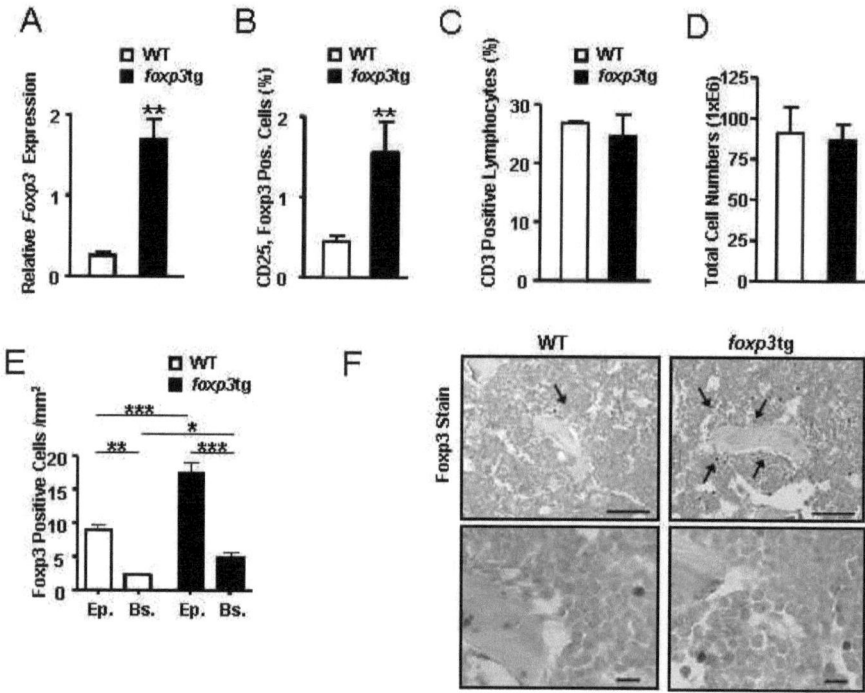

Figure 10. *Foxp3*tg mice show increased Treg numbers.

(A) Quantitative RT-PCR analysis of *Foxp3* expression in bone marrow cells. (B) Flow cytometry analysis of $CD25^+$ $Foxp3^+$ positive bone marrow cells ($n=6$). (C) Flow cytometry analysis of $CD3^+$ lymphocytes and (D) total splenic cell number counts. (E) Histochemical comparison of $Foxp3^+$ cells in tibia of WT and *foxp3*tg mice, measured in the epiphysis (Ep.) and in the bone shaft (Bs.). (F) Immunohistological sections from the epiphysis of tibia stained for Foxp3. Arrows indicate $Foxp3^+$ cells. Scale bars, 100μm.

Results

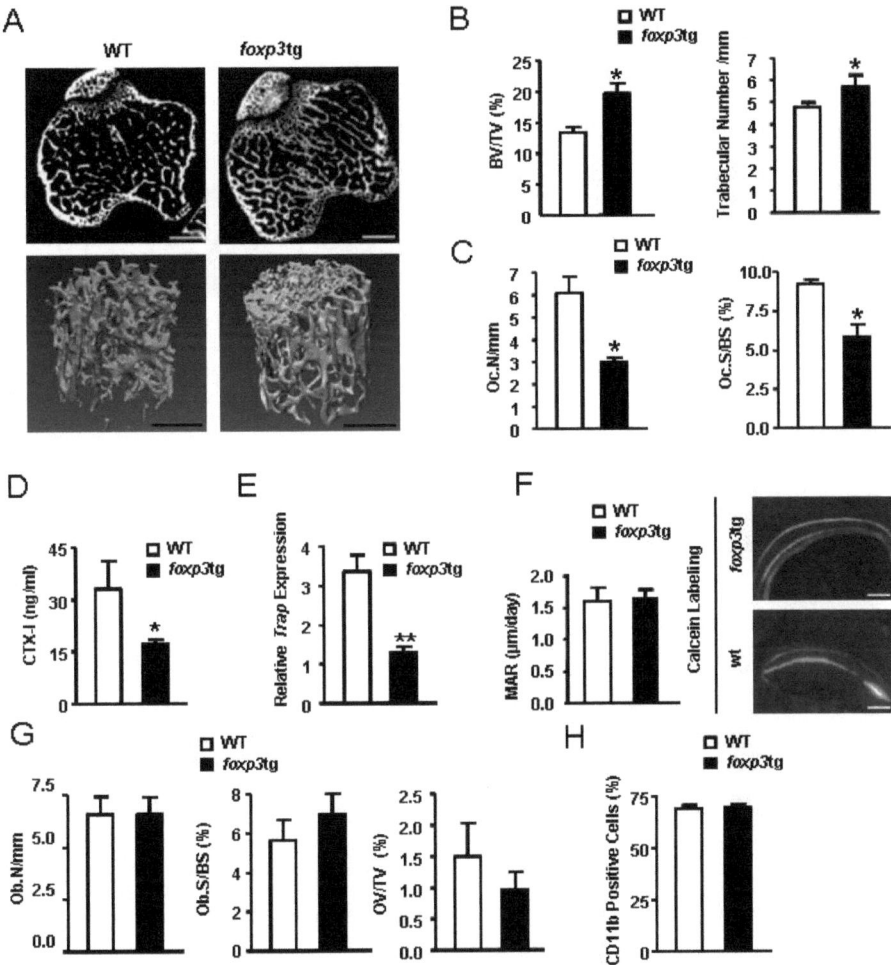

Figure 11. Increased bone density in *foxp3*-transgenic (*foxp3*tg) mice due to decreased bone resorption.

(A) Micro-computed tomography of trabecular bone from tibia of wild-type (WT) and *foxp3*tg mice. Scale bar, 0.25 mm. (B) Structural parameters of tibia from males (*n*=7) at week 12: bone volume/total volume (BV/TV) and trabecular numbers. (C) Quantitative histomorphometry of osteoclast numbers normalized to trabecular bone perimeter (Oc.N/mm) and osteoclast surface normalized to bone surface (Oc.S/BS). (D) Ratlaps™ (CTX-I) ELISA analysis for marker of bone destruction (*n*=5) and RANKL, OPG serum levels. (E) Quantitaive RT-PCR analysis of *Trap* expression in tibia. (F) Mineral apposition rate (MAR) in trabecular bone analyzed by calcein labeling (*n*=6). Scale bars, 2.5 µm. (G) Quantitative histomorphometry of osteoblast numbers normalized to trabecular bone perimeter (Ob.N/mm), osteoblast surface normalized to bone surface (Ob.S/BS) and ratio osteoid volume/total volume (OV/TV). (H) Flow cytometry analysis of osteoclast progenitors (CD11b[+] cells) in bone marrow cells gated on cells with high FCS and SSC in dot plots.

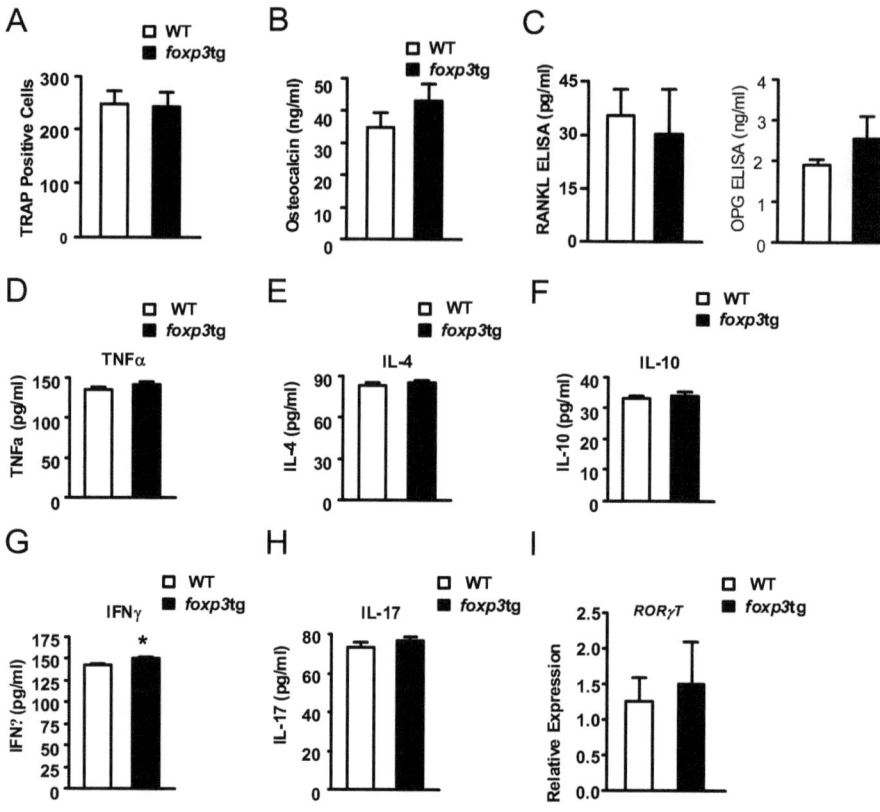

Figure 12. Osteoclast differentiation assay and serum parameters of *foxp3*tg mice.

(A) Osteoclast differentiation assay *in vitro* of WT and *foxp3*tg bone marrow cells. One out of 10 experiments is shown. **(B)** Osteocalcin ELISA analysis of bone formation marker in serum. Serum levels in *foxp3*tg mice compared to littermate controls (WT) measured with Cytoplex FACS analysis for **(D)** TNFα, **(E)** IL-4, **(F)** IL-10, **(G)** IFNγ and **(H)** IL-17. **(I)** Quantitaive RT-PCR analysis of *RORγT* expression in tibia.

To test the hypothesis that *in vivo*, increasing Treg numbers could locally affect osteoclast differentiation we compared the suppressive potential of splenic T cells of WT and *foxp3*tg mice in co-cultures with osteoclast precursors isolated from WT mice. Low concentrations of CD4+ T cells from *foxp3*tg mice suppressed osteoclastogenesis, whereas CD4+ T cells from WT mice were only effective at high concentrations **(Figure 13A)**. This effect was not caused by an enhanced suppressive potential of individual Tregs but rather based on the higher fraction of Tregs in *foxp3*tg mice as indicated by identical suppressive effects of WT and *foxp3*tg purified Tregs on osteoclast differentiation **(Figure 13B)**.

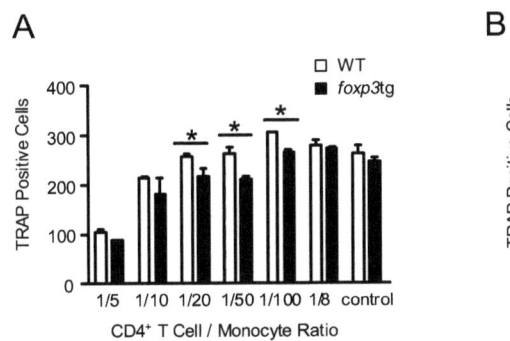
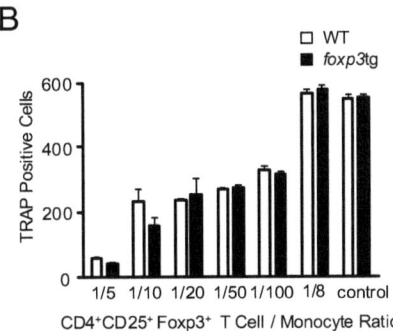

Figure 13. Tregs as the T cell subset that is responsible for suppression of osteoclast differentiation.

In vitro osteoclast differentiation of **(A)** CD4$^+$ T cells, **(B)** CD4$^+$CD25$^+$foxp3$^+$ T cells (Tregs) from wild-type (WT) and foxp3tg mice.

The numbers of Foxp3-positive regulatory T cells in peripheral blood of healthy human as well as in rheumatoid arthritis patients was inversely related to serum parameters of osteoclastogenesis, such as TRAP5b (Spearman r=-0.7311, P<0.0001), and bone resorption, such as collagen type 1 cleavage products (CTX-I) (Spearman r=-0.65, P<0.0001) **(Figure 14A and B)**. These data indicate that Tregs exert a suppressive effect on osteoclast formation and bone resorption in humans as well. In line with these results, no inverse correlation with parameters of bone formation, like osteocalcin, could be observed **(Figure 14C)**.

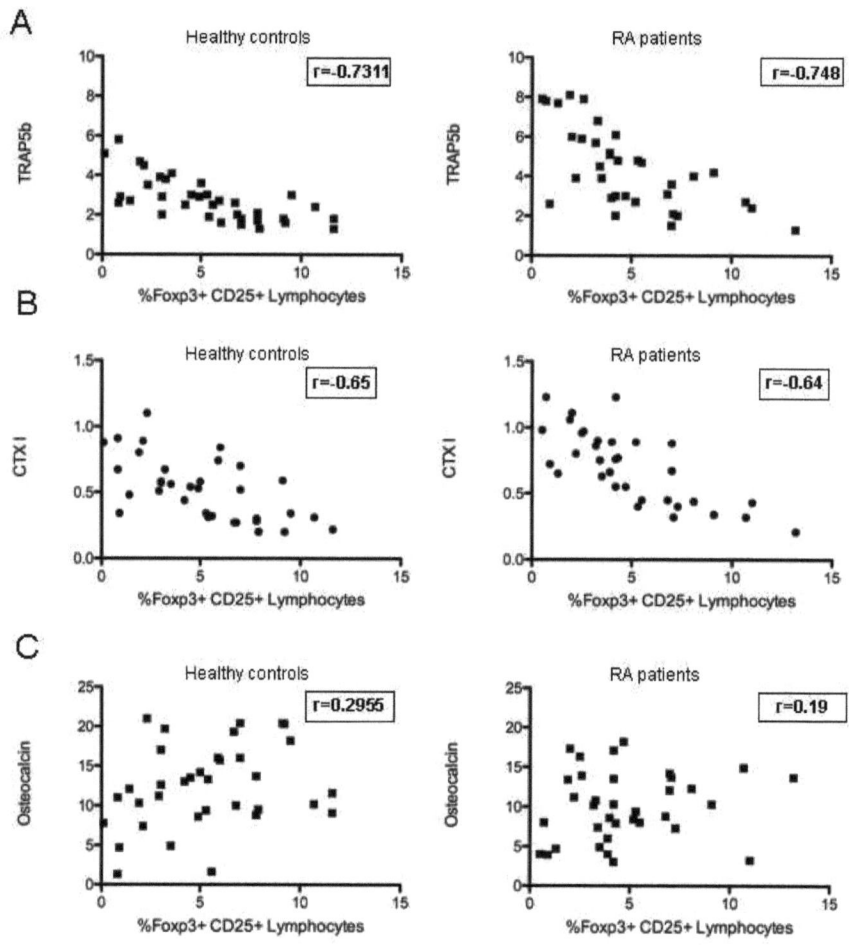

Figure 14. CD25⁺Foxp3⁺ T cells in peripheral blood of normal humans subjects (*n*=35) and RA patients (n=35) are inversely related to serum parameters of osteoclastogenesis.

(A) TRAP5b ELISA analysis correlated with flow cytometry analysis of CD25⁺Foxp3⁺ T cells in healthy controls (Spearman r=-0.7311, P<0.0001) and RA patients (r=-0.748, P<0.0001), or **(B)** collagen type 1 cleavage products of healthy controls (CTX-I) (Spearman r=-0.65, P<0.0001) and RA patients (Spearman r=-0.65, P<0.0001), and **(C)** osteocalcin (Spearman r=0.2955).

In summary, we could show that *foxp3tg* mice, with an increased fraction of Tregs, develop an increased bone density. This results from the decreased numbers of bone resorbing mature osteoclasts in these mice. This effect is based on an increase in Treg numbers and not due to an intrinsic defect of osteoclast precursor proliferation or a decrease in osteoblast numbers or activity.

3.3. Tregs and bone pathology

Given a role of Tregs in physiological bone homeostasis, we next questioned whether Tregs influence pathological bone remodeling.

3.3.1. *Foxp3tg* mice are protected from ovariectomy-induced bone loss

One of the major bone pathology affecting human is the development of osteoporosis caused by the decreased level of circulating estrogen following menopause. Indeed, increased osteoclast differentiation and activity not compensated by an increased bone formation is observed in more of one third of the aging women leading to osteoporosis. These situations can experimentally be mimicked by performing ovariectomy in mice. We thus performed an ovariectomy model in WT and *foxp3tg* mice to address whether Tregs could prevent postmenopausal bone loss. Ovariectomy led to a significant decreased bone mass (BV/TV: bone volume per total volume) of WT mice as observed by μCT analyses, whereas *foxp3tg* mice were significantly less affected **(Figure 15A and B)**. Bone mass of ovariectomized *foxp3tg* mice remained at the level of sham-operated wild-type mice **(Figure 15A and B)**. Moreover, the number of bone trabeculae was virtually maintained in *foxp3tg* mice after ovariectomy, but was strongly diminished and the major contributor to bone loss

Results

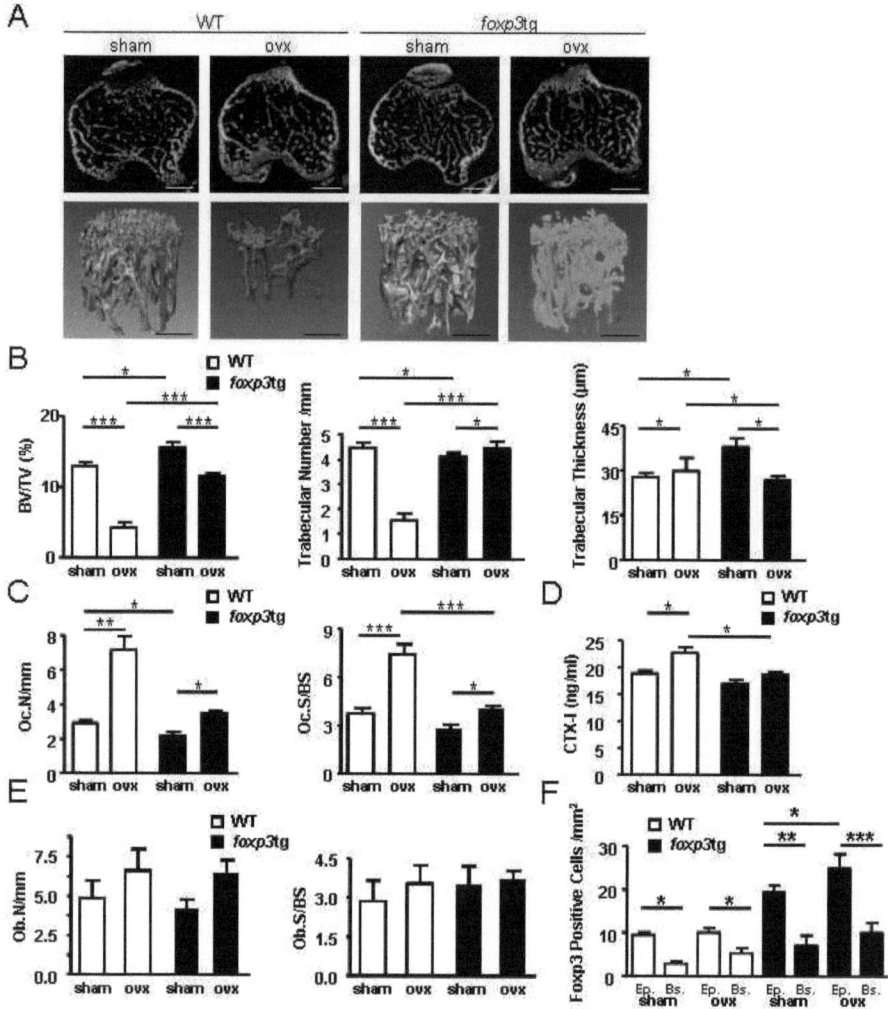

Figure 15. Foxp3tg mice are protected from bone loss after ovariectomy.

(A) Microcomputed tomography slice images and three-dimensional reconstruction of the trabecular bone from tibia of wild-type (WT) and *foxp3*tg mice either sham (sham) operated or ovariectomized (ovx). Scale bars, 0.25mm. **(B)** Structural parameters of trabecular bone from tibia of WT and *foxp3*tg female mice at week 12: bone volume/total volume (BV/TV), trabecular number and trabecular thickness. **(C)** Quantitative histomorphometry of osteoclast numbers normalized to trabecular bone perimeter (Oc.N/mm) and osteoclast surface normalized to bone surface (Oc.S/BS). **(D)** Ratlaps™ (CTX-I) ELISA analysis of bone destruction marker in serum (n=5). **(E)** Quantitative histomorphometry of osteoblast numbers normalized to trabecular bone perimeter (Ob.N/mm), osteoblast surface normalized to bone surface (Ob.S/BS). **(F)** Histochemical comparison of Foxp3$^+$ cells in tibia of WT and *foxp3*tg mice analysed in the epiphysis (Ep.) and in the bone shaft (Bs.).

in ovariectomized WT mice **(Figure 15A and B)**. Protection of *foxp3*tg mice from ovariectomy-induced bone loss was achieved by preventing enhanced osteoclast-mediated bone resorption, whereas a significant increase in osteoclast numbers and osteoclast-covered bone surface was observed in WT mice **(Figure 15C)**. The serum marker for bone resorption (CTX-I) remained low in *foxp3*tg mice, but increased significantly in WT mice after ovariectomy **(Figure 15D)** suggesting that Tregs control enhanced osteoclastogenesis in the postmenopausal state. Parameters of bone formation such as osteoblast numbers and osteoblast covered bone surface **(Figure 15E)** did not differ among *foxp3*tg and WT mice. Importantly, immunohistochemical analyses of tibia revealed an increased accumulation of Foxp3-expressing lymphocytes next to bony trabeculae in the epiphysis of both sham operated and ovariectomized *foxp3*tg mice **(Figure 15F)**.

3.3.2. *Foxp3*tg bone marrow protects from inflammation-induced bone loss

Bone destruction is also observed in patients suffering from rheumatoid arthritis. Indeed, local bone destruction is observed in inflamed joints of RA patients who also suffer from systemic bone loss. Tumour necrosis factor α (TNFα) plays an essential role in chronic inflammatory diseases and links inflammation to degradation of bone by stimulating osteoclastogenesis (105). The pathology is recapitulated in transgenic mice for human TNFα (*hTNF*tg), which develop destructive arthritis and severe osteopenia (100, 106). To address the role of Tregs in inflammatory bone loss we reconstituted *hTNF*tg mice with bone marrow from WT, *foxp3* mutant scurfy (*sf*/Y) or *foxp3*tg mice. Reconstitution with bone marrow from *sf*/Y mice significantly aggravated the osteopenic phenotype of *hTNF*tg mice **(Figure 16A and B)**. In contrast, *hTNF*tg/*foxp3*tg chimeras maintained a higher bone mass and were resistant to systemic bone loss **(Figure 16A and B)**, suggesting that the activity of the Tregs pool is a key player in skeletal homeostasis during inflammation. Structural analyses showed that regulation of bone mass is caused by a significantly higher trabecular number in *hTNF*tg/*foxp3*tg and lower trabecular numbers in *hTNF*tg/*sf*/Y chimeras than in *hTNF*tg/wt chimeras while trabecular thickness was unaffected **(Figure 16B)**.

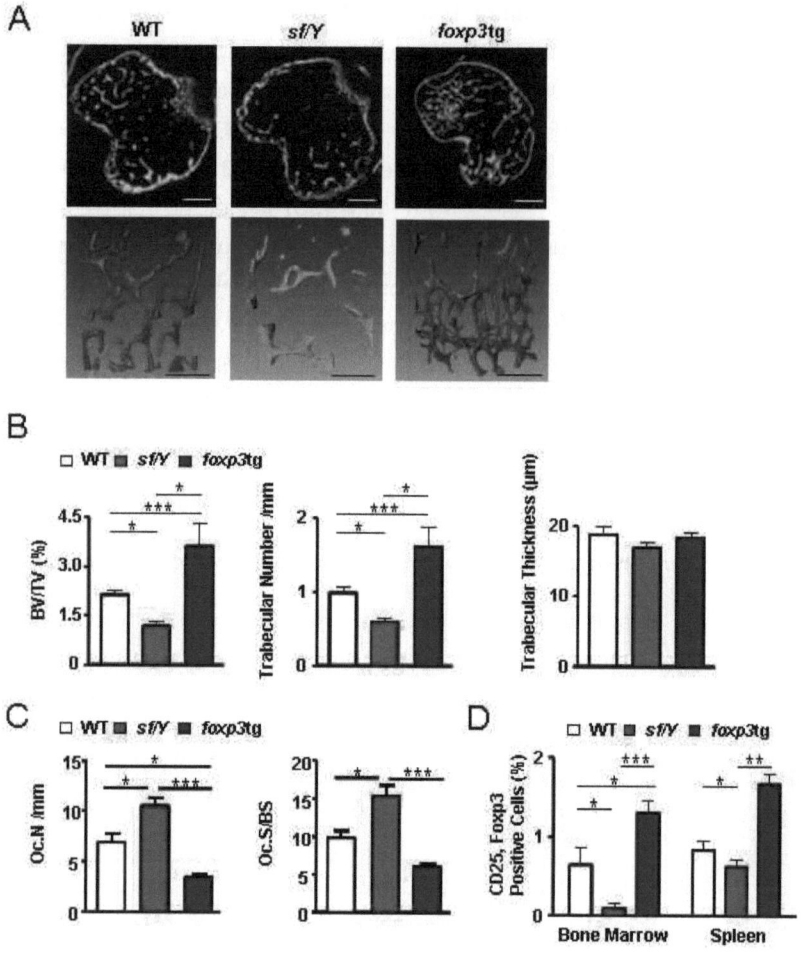

Figure 16. Foxp3tg bone marrow protects hTNF-transgenic (hTNFtg) mice -an arthritis mouse model- from systemic bone loss.

(A) Micro-computed tomography slice images and three-dimensional reconstruction of trabecular bone from irradiated hTNFtg mice reconstituted with bone marrow from wild-type (WT), scurfy (sf/Y) or foxp3tg mice. Scale bars, 0.25mm. (B) Structural parameters of trabecular bone from tibia of bone marrow reconstituted hTNFtg mice at week 12. Bone volume/total volume (BV/TV), trabecular numbers and trabecular thickness. (C) Quantitative histomorphometry of osteoclast numbers normalized to trabecular bone perimeter (Oc.N/mm) and osteoclast surface normalized to bone surface (Oc.S/BS). (D) Flow cytometry analysis of $CD25^+Foxp3^+$ cells from bone marrow and spleen cells (n=5).

Osteoclast parameters such as osteoclasts numbers and osteoclast-covered bone surface were significantly decreased in *hTNFtg/foxp3*tg chimeras whereas a significant increase was observed in *hTNFtg/sf/Y* chimeras **(Figure 16C)**. In addition, flow cytometry analyses in bone marrow and spleen cells from TNFtg chimera animals revealed and significant higher cell fraction of Tregs in *hTNFtg/foxp3*tg animals compared to *hTNFtg/sf/Y* and *hTNFtg*/wt animals **(Figure 16D)**.

To determine if Tregs may also influence the destruction of joint architecture in inflammatory arthritis, we analyzed periarticular bone and cartilage in *hTNFtg* mice challenged with bone marrow from WT, *sf/Y* and *foxp3*tg mice. Progressive clinical signs of arthritis such as paw swelling, loss of grip strength and body weight were found in *hTNFtg*/wt chimeras **(Figure 17A)**. Disease course was more pronounced in *hTNFtg/sf/Y* chimeras, whereas *hTNFtg/foxp3*tg chimeras showed a significantly milder disease course. Histology revealed an almost complete preservation of joint architectures in *hTNFtg/foxp3*tg, whereas *hTNFtg/sf/Y* mice exhibited the most severe structural damages with dramatic bone loss and cartilage destruction **(Figure 17B and C)**. Osteoclast numbers in the synovium were higher in *hTNFtg/sf/Y* chimeras and drastically reduced in *hTNFtg/foxp3*tg chimeras, indicating that the suppressive effect of Tregs extends to local osteoclastogenesis in the joint **(Figure 17B and C)**. In histological sections a significant, conspicuous accumulation of Foxp3 expressing T cells especially at sites of bone destruction was observed **(Figure 18A)**. This suggests that Tregs exert their anti-osteoclastogenic effect locally at sites where osteoclasts are formed. Direct *in vitro* osteoclast assays from *hTNFtg/foxp3*tg and *hTNFtg*/wt chimeras showed no difference confirming no intrinsic change of osteoclastogenic potential of mononuclear cells **(Figure 18B)**. Moreover, osteoclast formation in *hTNFtg/sf/Y* chimeras was higher, but this effect was based on altered redistribution of CD11b$^+$ osteoclast precursors with higher numbers in the bone marrow and spleen of these mice **(Figure 18C)**.

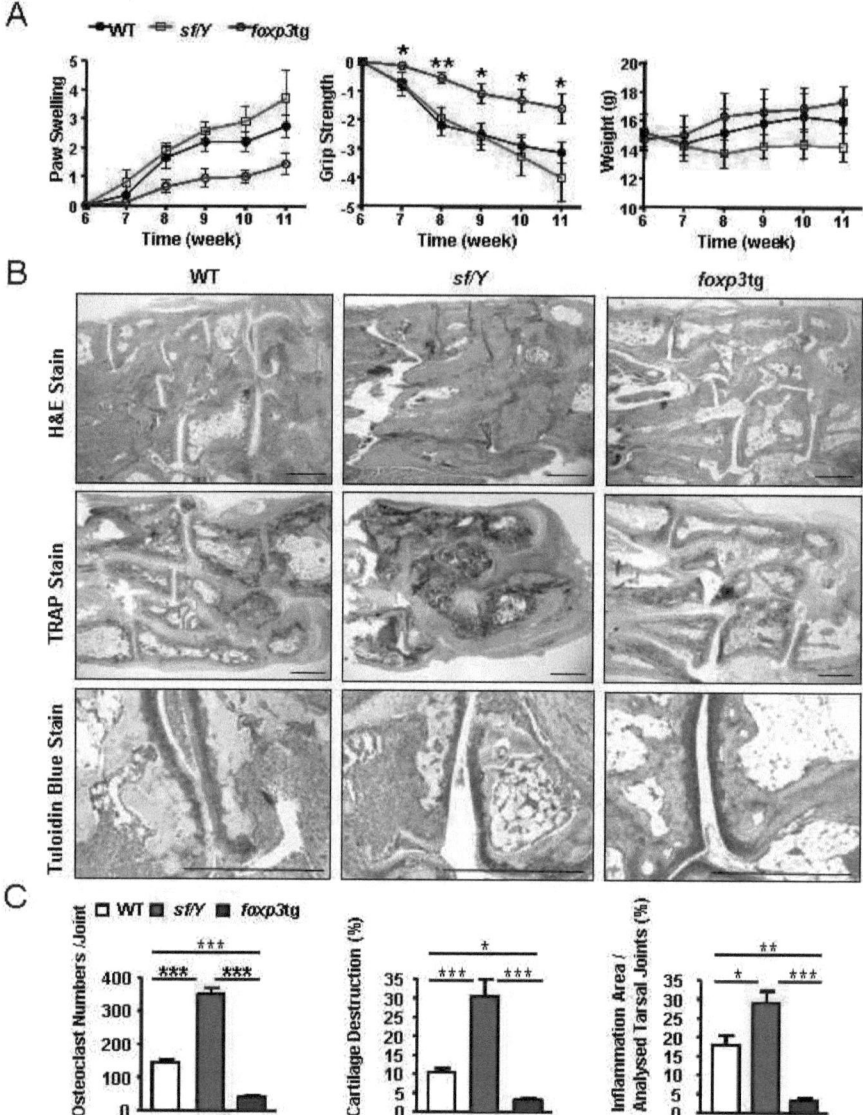

Figure 17. **Foxp3tg bone marrow protects hTNF-transgenic (hTNFtg) mice from inflammation-induced bone destruction.**

(A) Clinical parameters from irradiated hTNFtg mice reconstituted with bone marrow from wild-type (WT), scurfy (sf/Y) or foxp3tg mice. Paw swelling, grip strength and body weight measurements. (B) Histological sections of tarsal joints from hTNFtg mice after bone marrow transfer of WT, sf/Y or foxp3tg mice stained with H&E (high panel), TRAP (middle panel) and Tuloidin Blue (low panel). Scale bars, 500μm. (C) Histomorphometric assessment of osteoclast numbers, cartilage destruction and synovial inflammation in tarsal joints.

Results

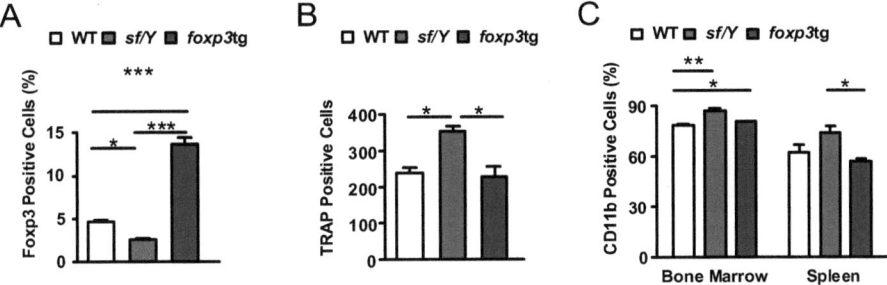

Figure 18. h*TNF*-transgenic (h*TNF*tg) bone marrow chimera mice have no intrinsic change of osteoclast differentiation *in vitro*.

(A) Immunohistological analysis of Foxp3⁺ T cells at the side of bone destruction in tarsal joints. **(B)** Osteoclast differentiation assay of bone marrow cells *in vitro*. One out of 5 experiments is shown. **(C)** Flow cytometry analysis of osteoclast progenitors (CD11b⁺ cells) in bone marrow and spleen from irradiated *hTNF*tg mice reconstituted with bone marrow from wild-type (WT), scurfy (*sf/Y*) or *foxp3*tg mice gated on cells with high FCS and SSC in dot plots (*n*=5).

3.3.3. SupCD28Mab ameliorates TNF-induced arthritis and systemic bone loss

In addition to the bone marrow chimeras, we treated *hTNF*tg mice with a CD28 superagonist monoclonal antibody (supCD28Mab) for selective targeting and expansion of Tregs in the treated mice (107). Mice were injected twice during the study period and progressive clinical signs were analyzed **(Figure 19A)**. Disease course was ameliorated in *hTNF*tg mice treated with the supCD28Mab. These mice showed less paw swelling, better grip strength and an increased weight compared to untreated controls (control). Histological analyses of the tarsal joints in the paws showed significantly better joint architectures in treated *hTNF*tg mice **(Figure 19B)**. Treated mice had less osteoclast numbers, cartilage destruction and inflamed tissue in the synovium of tarsal joints **(Figure 19B)**. In addition to local inflammation induced bone destruction in the paws, supCD28Mab treated *hTNF*tg mice showed also and effect on systemic bone density **(Figure 19C)**. Systemic bone density was significantly increased in treated mice, which results mainly from an increase in trabecular numbers in these mice **(Figure 19C)**. Both, preservation of local joint architecture in the paw as well as systemic bone density indicates to an protective role of Tregs in supCD28Mab treated mice, since flow cytometry analyses of spleen and bone marrow cells revealed a significant increase in CD4⁺Foxp3⁺ Tregs in treated mice **(Figure 19D and E)**.

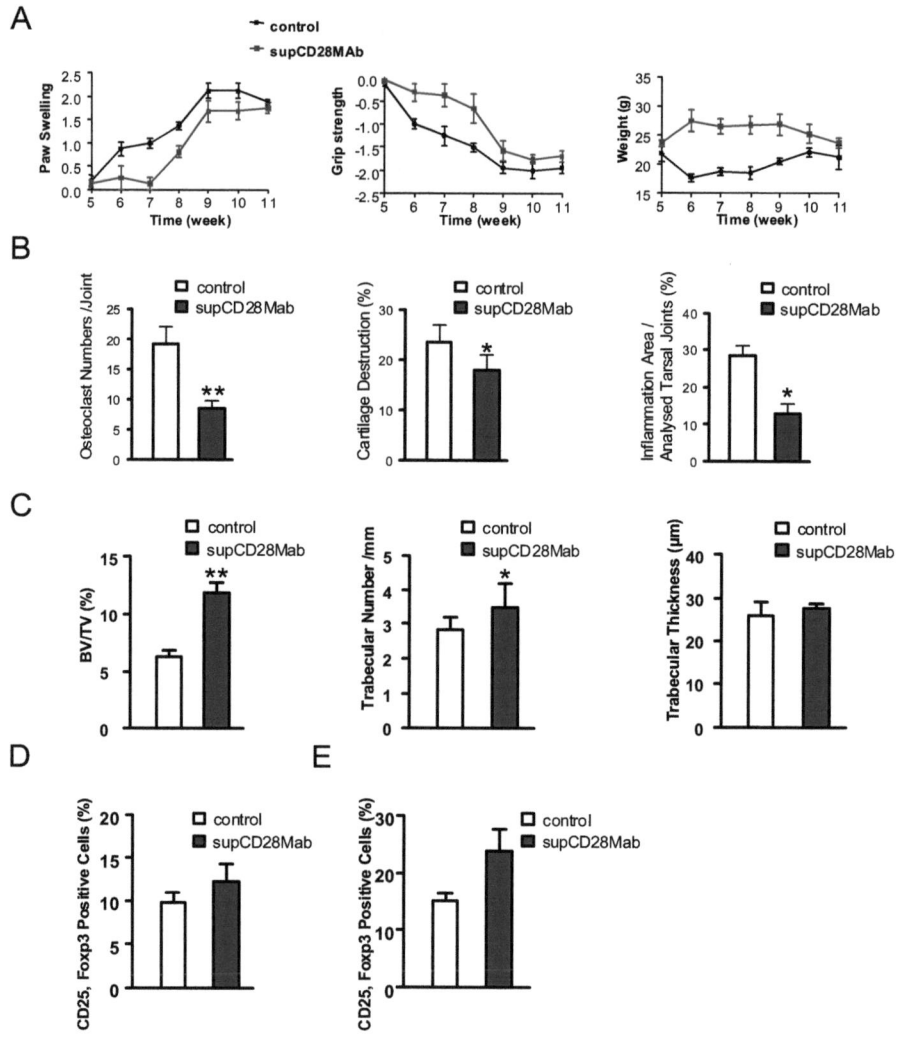

Figure 19. SupCD28Mab ameliorates arthritis and systemic bone density in *hTNF*tg mice.

(A) Clinical parameters from h*TNF*tg mice treated twice with supCD28Mab within 6 weeks. Paw swelling, grip strength and body weight measurements. **(B)** Histomorphometric assessment of osteoclast numbers, cartilage destruction and synovial inflammation in tarsal joints. **(C)** Structural parameters of trabecular bone from tibia of supCD28Mab treated *hTNF*tg mice at week 12: bone volume/total volume (BV/TV), trabecular number and trabecular thickness. Flow cytometry analysis of $CD25^+Foxp3^+$ cells in spleen **(D)** and bone marrow cells **(E)**.

In summary, we could show that *foxp3*tg mice show a significantly milder from of osteoporosis with less bone loss in an ovariectomy model. In addition, h*TNF*tg mice, a model for inflammatory arthritis, are protected from both inflammation induced bone loss in arthritic paws as well as from systemic bone loss after bone marrow transfer from *foxp3*tg mice and after expansion of the Treg population with supCD28Mab treatment. In both pathological models we observed a decreased bone resorption due to less osteoclast numbers.

3.4. Mechanism of osteoclast suppression by Tregs

3.4.1. Increased osteoclast numbers in *CD80/86$^{-/-}$* mice

We demonstrated *in vitro* the key role played by CTLA-4 interaction with CD80/86 in the suppression of osteoclast differentiation by Tregs. We thus analyzed the bone of CD80/86 deficient mice. µCT analyses of bone of *CD80/86$^{-/-}$* mice revealed significant osteopenia as shown by lower bone volume and trabecular numbers **(Figure 20A and B)**. Importantly, bone phenotype of *CD80/86$^{-/-}$* mice was based on increased osteoclast numbers and osteoclast covered bone surface as determined by histological analyses of trabecular bone **(Figure 20C and D)**. In contrast, bone structure was normal in *CD28$^{-/-}$*, *ICOSL$^{-/-}$* or *ICOS$^{-/-}$* mice (Figure 18E) and osteoclast counts did not differ from WT mice **(Figure 20F)**. A slight albeit not significant decrease in BV/TV and an increase in osteoclast numbers and osteoclast surface per bone surface (Oc.S/B.s.) in *CD28$^{-/-}$* mice were found **(Figure 20F)**. These data illustrate that the molecular mechanism by which Tregs protect bone is based on CD80/86, a key regulator of physiological bone homeostasis.

3.4.2. IDO is essential for CTLA-4-mediated osteoclast suppression

How does the interaction of CTLA-4 on Tregs with CD80/86 on osteoclast precursors regulate osteoclastogenesis? CD80/86 has been shown to regulate the induction of indoleamine 2,3 dioxygenase (IDO), a catabolic enzyme responsible for degradation of the essential amino acid tryptophan along the kynurenine pathway, in dendritic cells (108). We tested a potential involvement of IDO in the suppressive effect of Tregs on osteoclastogenesis. In our experiments IDO mRNA expression was induced

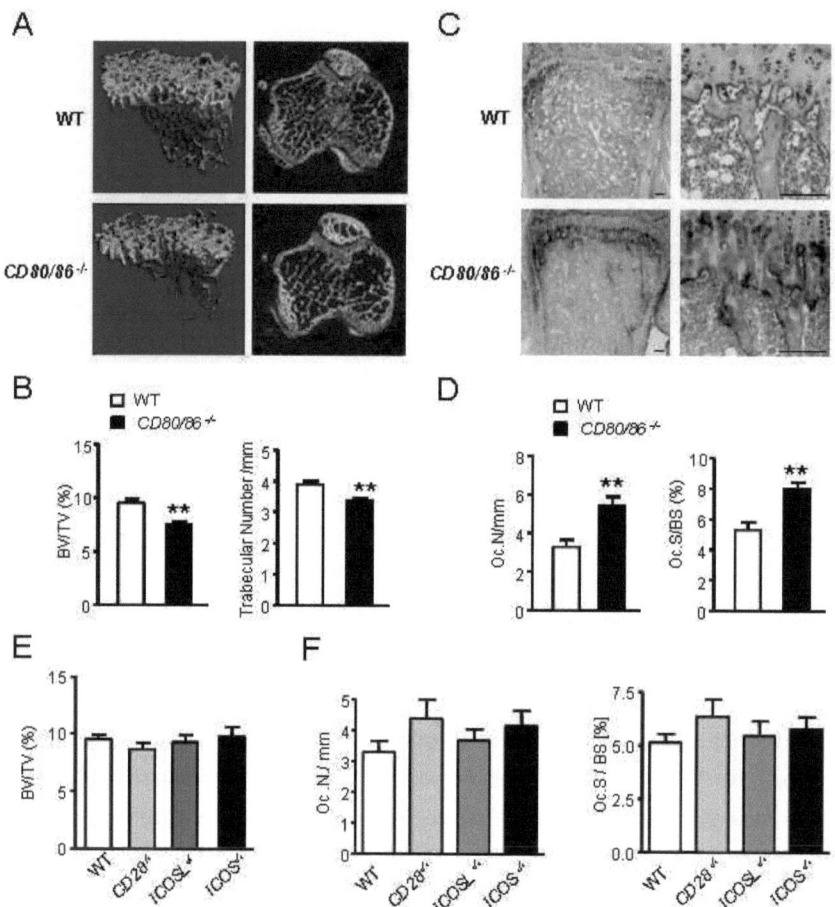

Figure 20. Decreased bone density in CD80/86 $^{-/-}$ mice.

(A) Micro-computed tomography of trabecular bone from tibia of wild type (WT) and CD80/86 $^{-/-}$ mice. **(B)** Structural parameters of male tibia (n=6) at week 12: Bone volume/total volume (BV/TV), trabecular number. **(C)** TRAP stained sections of tibia from WT and CD80/86 $^{-/-}$ mice. Scale bars, 100 µm. **(D)** Quantitative histomorphometry of osteoclast numbers normalized to trabecular bone perimeter (Oc.N/mm) and osteoclast surface normalized to bone surface (Oc.S/BS). **(E)** Structural parameters of trabecular bone from tibia of WT, CD28 $^{-/-}$, ICOSL $^{-/-}$ and ICOS $^{-/-}$: Bone volume/total volume (BV/TV). **(F)** Quantitative histomorphometry of osteoclast numbers normalized to trabecular bone perimeter (Oc.N/mm) and osteoclast surface normalized to bone surface (Oc.S/BS).

in CD11b⁺ monocytes of co-cultures with Tregs **(Figure 21A)**. The activity of IDO can be detected by measuring kynurenine / tryptophan ratio. Induced IDO mRNA expression also resulted in an increased kynurenine/tryptophan ratio in supernatants of co-cultures **(Figure 21B)**. Similarly, IDO expression was up-regulated by treatment of monocytes with CTLA-4 Ig **(Figure 21D)** also resulting in an increased kynurenine/tryptophan ratio **(Figure 21E)**. Western blot analyses confirmed the upregulation of IDO after CTLA-4 Ig treatment **(Figure 21G)**. Similar results were obtained in co-culture experiments with Tregs **(Figure 21I)**. The IDO upregulation in CD11b⁺ monocytes led to a significant increased apoptosis frequency of the osteoclast precursors **(Figure 21C and F)**. This increased apoptosis was due to the upregulation of IDO expression in CD11b⁺ monocytes after binding of Tregs via surface expression of CTLA-4 or CTLA-4 Ig alone with CD80/86 on CD11b⁺ monocytes. This was confirmed by the lack of (i) IDO mRNA **(Figure 21D)** or (ii) protein expression **(Figure 21G)** and (iii) the resistance to apoptosis of CD11b⁺ monocytes isolated from *CD80/86⁻/⁻* mice co-cultured with Tregs or in presence of CTLA-4 Ig **(Figure 21C and F)**. We thus performed specific knock-down experiments to analyze the role of IDO in osteoclastogenesis. When CD11b⁺ osteoclast precursors were transfected with siRNAs for IDO, the suppressive effect of CTLA-4 on Tregs was specifically abolished as indicated by an significant increased number of TRAP positive cell in co-culture experiments with Tregs **(Figure 21G)**. The efficiency, especially for IDO siRNA3, out of the 3 different used IDO siRNAs is shown in western blot analyses **(Figure 21H)**. These data show that CTLA-4 engaging CD80/86 induces IDO expression in osteoclast precursors and mediates the suppressive effects of Tregs on osteoclast differentiation.

Results

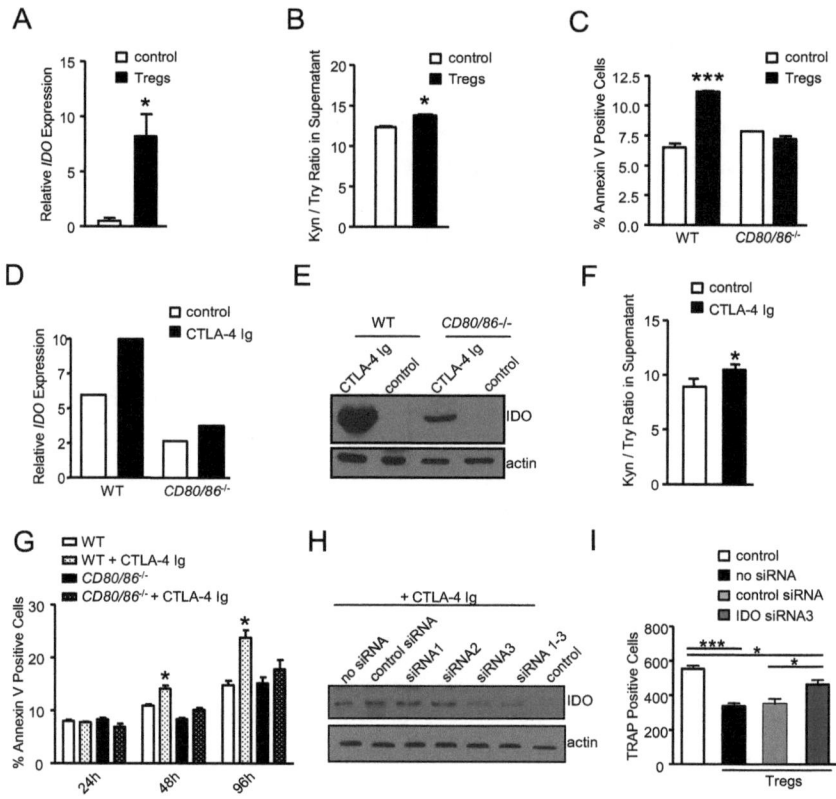

Figure 21. Induction of IDO-mediated Tryptophan catabolism is essential for osteoclastogenesis inhibition by Tregs.

(A) Quantitative RT-PCR analysis of *IDO* expression in monocytes from co-cultures of monocytes in the absence (control) or in the presence of Tregs. **(B)** IDO activity determined by measuring the ratio kynurenine/tryptophan (Kyn/Try) in the supernatant of the same condition. **(C)** Apoptosis analysis by flow cytometry of monocytes from WT and $CD80/86^{-/-}$ mice cultured in osteoclastogenic conditions either in the absence (control) or in the presence of Tregs. **(D)** Quantitative RT-PCR analysis of *IDO* expression in monocytes stimulated by CTLA-4 Ig compared to isotype control stimulated cells (control) in WT and $CD80/86^{-/-}$ mice. **(E)** Western Blot analysis of same experimental setting as in (D). **(F)** Kynurenine/tryptophan (Kyn/ Try) ratio in the supernatant of CTLA-4 Ig stimulated WT monocytes. **(G)** Apoptosis analysis by flow cytometry of monocytes from WT and $CD80/86^{-/-}$ mice cultured in osteoclastogenic conditions stimulated by CTLA-4 Ig compared to isotype control stimulated. **(H)** Western Blot analysis of monocytes after transfection with IDO siRNA or a non-specific siRNA (control siRNA) compared to untransfected cells (no siRNA) or without CTLA-4Ig stimulation (control). **(I)** Suppressive effect of Tregs on osteoclastogenesis following monocytes transfection with IDO siRNA or a non-specific siRNA (control siRNA) compared to untransfected cells (no siRNA).

3.4.3. Increased bone density in $RAG1^{-/-}$ mice after adoptive Treg transfer

To address whether the effect of Tregs on bone homeostasis in indirect through T cell suppression or by directly engaging osteoclast precursors, we performed an adoptive transfer of Treg model. We injected $1x10^6$ sorted $CD4^+CD25^+$ T cells into lymphocyte deficient RAG-1-deficient ($RAG1^{-/-}$) mice for 2 times within 6 weeks. Sorted $CD4^+CD25^+$ T cells were analysed by flow cytometry for Foxp3 expression as described in Figure 1, and more then 90% of injected $CD4^+CD25^+$ T cells were Foxp3$^+$ Tregs. 6 weeks after the first Treg transfer, mice were analyzed by flow cytometry and quantitative PCR for $CD4^+CD25^+Foxp3^+$ T cells. Flow cytometry and RT-PCR analyses of splenic cells from $RAG1^{-/-}$ mice (**Figure 22A and B**) showed detectable levels of Tregs in transferred mice compared to controls. Interestingly, serum levels of CTX-I was decreased in Treg- transferred $RAG1^{-/-}$ mice compared to control mice, whereas levels of osteocalcin were normal (**Figure 22D and E**). There was also an increase in both OPG and RANKL in Treg- challenged $RAG1^{-/-}$ mice compared to control mice (**Figure 22F and G**) but there was no significant influence on RANKL/OPG ratio. µCT analyses of tibias in Treg- treated and untreated sex-matched $RAG1^{-/-}$ mice revealed a significant increased bone mass (BV/TV: bone volume per total volume) in Treg- challenged mice (**Figure 22H**). The phenotype was based on higher number of bony trabeculae (**Figure 22H**). Decreased osteoclasts numbers and osteoclast-covered bone surface indicated an impaired bone resorption in treated $RAG1^{-/-}$ mice (**Figure 22G**), whereas there was no change in numbers of osteoblasts or osteoblast-cover bone surface (**Figure 22H**).

Results

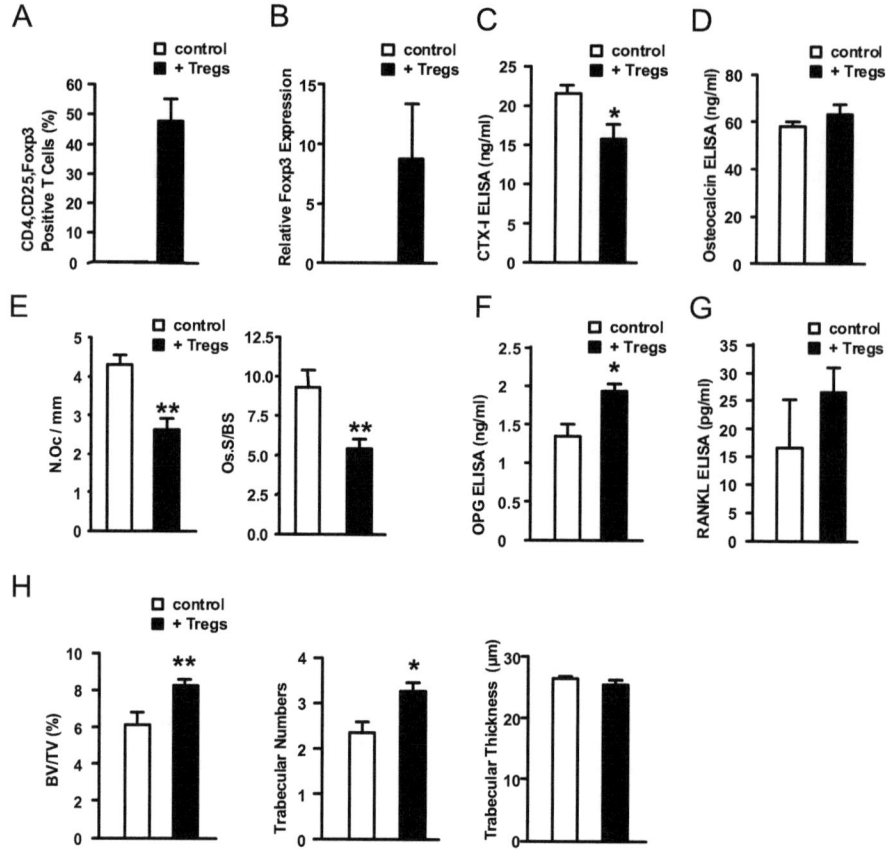

Figure 22. Increased bone density in lymphocyte deficient $RAG1^{-/-}$ mice after adoptive Treg transfer.

(A) Flow cytometry analysis of spleen cells from $RAG1^{-/-}$ mice with (+ Tregs) or without adoptive Treg transfer (control). **(B)** Quantitative RT-PCR analysis of *foxp3* expression in spleen cells from $RAG1^{-/-}$ mice with (+ Tregs) or without adoptive Treg transfer (control). **(C)** RatlapsTM (CTX-I) ELISA serum analysis for marker of bone destruction in $RAG1^{-/-}$ mice. **(D)** Osteocalcin ELISA serum analysis as marker for bone formation. **(E)** Quantitative histomorphometry of osteoclast numbers normalized to trabecular bone perimeter (Oc.N/mm) and osteoclast surface normalized to bone surface (Oc.S/BS). **(F)** OPG and **(G)** RANKL ELISA anaylsis of serum levelsin $RAG1^{-/-}$ mice. **(H)** Structural parameters of tibia from $RAG1^{-/-}$ mice at week 12: bone volume/total volume (BV/TV), trabecular numbers and trabecular thickness.

In summary, we showed that mice with increased Tregs numbers developed high bone mass and were protected from ovariectomy-induced osteoporosis, inflammatory osteopenia and arthritic bone destruction, whereas *foxp3*-deficiency enhanced bone loss. Tregs directly protect against bone loss since a significantly increased bone mass was observed after adoptive transfer of Tregs RAG1 deficient mice ($RAG1^{-/-}$). The positive skeletal effects of Tregs were all mediated through inhibition of bone resorption by osteoclasts. Binding of Tregs to CD80/CD86 on osteoclast precursors was essential for inhibiting osteoclast differentiation *in vitro*. In agreement, CD80/86 deficient mice ($CD80/86^{-/-}$) developed osteopenia due to increased osteoclast differentiation. Engagement of CD80/86 by Tregs through CTLA-4 induced IDO expression in osteoclast precursors and thereby increased tryptophan catabolism and apoptosis in osteoclast precursors. Importantly, CD80/86 deficient osteoclast precursors showed no increased IDO expression and did not become apoptotic after engagement with CTLA-4. These results demonstrate that $Foxp3^+$ Tregs can actively and directly control bone resorption during physiological and pathological bone remodeling and thus preserve bone mass.

4. Discussion

In this thesis, we unravel a new link between the immune system and bone that addresses the regulatory interaction between these two organ systems. We show that naturally occurring $CD4^+CD25^+Foxp3^+$ regulatory T cells (Tregs) suppress osteoclast formation in a cell contact–dependent manner *in vitro* and that direct interaction of CTLA-4 on Tregs and CD80/86 on monocytes as osteoclast precursors play and essential role. The physiological importance of these finding was underlined by the observation that Tregs protect against bone destruction in both normal and pathological conditions *in vivo* and thus uncover a novel bone protection mechanism by the immune system.

4.1. Tregs as potential negative regulators of bone remodeling

Physiological, bone remodeling is often seen as a controlled inflammatory reaction. Considering the intensive crosstalk during bone remodeling between cells of the immune system and bone cells, it is of key interest to describe the regulatory mechanisms linking bone resorption to cells that regulate the inflammatory responses in this process.

Inflammation and bone loss are two frequently occurring and tightly linked disorders. Uncontrolled inflammatory responses often lead to profound bone loss, which increases skeletal fragility. This is seen in various inflammatory diseases such as rheumatoid arthritis (RA), systemic lupus erythematosus (SLE), psoriasis and inflammatory bowel disease that are all characterized by dramatic skeletal damage, which contributes to a high burden of disease due to fractures, bone erosion, functional disability, and even crippling (109).

In line with this, inflammation and bone loss share several features. For example, tumor necrosis factor-alpha (TNFα) is mainly known to act on cells of the immune system but it exerts its effects also on bone cells. On osteoblasts, TNFα inhibits extracellular matrix deposition, stimulates matrix metalloproteinase synthesis (110), and also enhances production of osteoclastogenic cytokines such as M-CSF and RANKL (50, 111). TNFα also enhances bone resorption both *in vitro* and *in vivo* by stimulating the proliferation and differentiation of osteoclasts (112, 113). In addition, TNFα is one of the key regulatory molecules for the trafficking of osteoclast precursors to these sites of inflammation (114). Inflammatory cytokines secreted by cells of the immune system can induce osteoclast differentiation, and many of them act via inflammatory T cell subsets that drive osteoclast formation by the expression of M-CSF and RANKL (105). Last but not least cytokines that have regulatory properties in inflammatory responses, such as IL-4, and IL-10, also negatively regulate osteoclast formation, suggesting that the negative regulation of immune activation follows principles similar to those involved in the regulation of bone resorption (115).

In the literature, however, studies prevail which show the stimulatory effect on bone resorption by activated immune cells, mainly $CD4^+$ T cells (116, 117). Since cells of the immune system stimulate bone resorption it is likely that a specific population of immune cells might also exert suppressive effects on bone loss. For several reasons, Tregs appear to be an attractive candidate that could represent this suppressive link between the immune system and bone. First, Tregs are the best-defined source for balancing immune activation, and alterations in function or number of these cells leads to severe inflammatory diseases such as Crohn's Disease (118), systematic lupus (SLE) (119), diabetes mellitus type I (120), autoimmune Hepatitis (121) and rheumatoid arthritis (122-124). In mutant mice, which lacks Tregs (scurfy mice) we can observe similar changes. The dramatic phenotype observed in scurfy mice appears to result from an inability to appropriately regulate T cell function (101). This phenotype is comparable to CTLA-4 (125) or TGF-β (126) knock out mice which also suffer from a severe and rapidly fatal lympho-proliferative disease resulting in death within 3 weeks of age. Second, as members of the T cells lineage, Tregs can engage with osteoclast precursors and mimic the interaction between activated T cells and antigen presenting cells (APC). The interaction with APCs is an important suppressive mechanism. Indeed recent studies suggested that Tregs act on cells of the innate immune system in order to control arthritis (127, 128). For the interaction of Tregs with APCs binding of CTLA-4 to CD80 and CD86 on APCs is important. It was shown that binding of Tregs to APCs, especially DCs, abrogates the antigen-presenting activity of DCs and down-regulates the costimularory molecules on DCs, which are key mediators to T cell activation (129, 130). Finally, Tregs express cytokines such as TGFβ, IL-4, and IL-10, which not only have anti inflammatory properties but also inhibit osteoclast formation. Although it was shown that TGFβ induce osteoclast differentiation by down-regulation of OPG in synovial fibroblasts from RA patients, the opposite effect is found in synovial fibroblasts from osteoarthritis (OA) patients (131), indicating that TGFβ could either suppress or enhance osteoclast differentiation. For the other cytokines expressed by Tregs, studies have shown that they also have suppressive effects on osteoclast differentiation. For example IL-4, induces OPG expression through activation of STAT6 in endothelial cells, which results in the suppression of osteoclast differentiation (132). In addition to IL-4, IL-10 also directly suppresses osteoclast differentiation through inhibition of RANKL-mediated expression of NFATc1 (133).

These were the reasons why we investigated the role of Tregs in bone remodeling. The results from our initial *in vitro* co-culture experiments show a dose dependent suppression of Tregs on osteoclast differentiation. Osteoclast differentiation was inhibited after Tregs bind osteoclast precursors via interaction between CTLA-4 and CD80/86. This finding nicely fits to already published suppressive properties of Tregs on T cells.

4.2. Mechanisms of osteoclast suppression by Tregs

We found that Tregs can effectively suppress osteoclast differentiation. This is in clear contrast to the role of inflammatory T cell subsets in stimulating osteoclast formation (86, 89). Although naturally occurring $CD4^+CD25^+Foxp3^+$ Tregs express RANKL, our results clearly show that the suppression of osteoclast formation occurs in a cell contact–dependent manner (99). We and others have reported that Tregs exert a suppressive effect of on osteoclastogenesis (97-99), but with different emphasis on potential suppressive mechanisms. Other *in vitro* data obtained from the differentiation of human monocytes into bone resorbing mature osteoclasts suggested that secretion of cytokines like IL-4, IFNγ and IL-10 was essential for the anti-osteoclastogenic effect on human Tregs (98). Another group reported that IL-5 and TGFβ suppress osteoclast differentiation in mice (97). Contrarily, we observed that blocking the receptors for IL-4, IL-10 and TGFβ on osteoclast precursors, only partially reverted the suppressive effects of Tregs on osteoclast differentiation. Thus we suggest that regulatory cytokines expressed by Tregs participate to the suppressive effect of Tregs on osteoclasts. But, as the suppressive effect has not been observed in Transwell experiments in our study, in which cells were mechanically separated, we think that direct cell-contact is required for suppression of osteoclastogenesis by Tregs suggesting that the secreted cytokines only have modulatory effects. The differences between our results and the results from the other two groups (97, 98) might be explained by different experimental settings. The main differences were (i) the source of osteoclast precursors, (ii) the concentrations of osteoclast stimulating cytokines and (iii) the number of Tregs in the co-cultures. First, we used sorted $CD11b^+$ monocytes from bone marrow cells as osteoclast precursors and not complete PBMCs. With this experimental setting we were able to reduce the contamination with other cells, such as fibroblasts, macrophages or DCs in the co-culture with Tregs, since these cells are an additional source for the suppressive cytokines and could contribute to the observed suppressive effect in cultures. Second, osteoclast precursors were challenged with twice higher concentrations of RANKL and thus may overcome the suppressive effects through over-stimulation. And third, in the Transwell experiments in the other studies a 4-fold higher Treg numbers was used that result in higher cytokine concentrations in culture media, which could again directly act on osteoclast precursors or indirectly stimulate cells out of the unsorted osteoclast precursor pool in the lower chamber to contribute to the suppressive effect. Supporting the need for cell-contact, the *in vitro* suppressive effect of Tregs is not based on a specific change in RANKL expression or OPG production by these cells, suggesting that interference with the RANKL–RANK interaction is not the cause of inhibition of osteoclast formation. The molecular mechanism of suppression is based on CTLA-4 (CD152), which is constitutively expressed by Tregs (70, 75). CTLA-4 impairs osteoclast formation by binding to osteoclast precursors and inhibiting their differentiation in a dose-dependent manner (99, 134). In agreement with these two studies, neutralization of CTLA-4 or CD80/86 restored the suppressive effect on osteoclast differentiation. The key role of CD80/86 was confirmed using monocytes from CD80/CD86 double knockout mice (*$CD80/86^{-/-}$*) that display a significant protection from suppression of Tregs. The fact that a single Treg cell is sufficient to block differentiation of about 100 monocytic cells into the osteoclast lineage underlines the powerful role of these cells in the regulation of bone resorption.

Considering the origin of osteoclasts from the mononuclear cell linage and their close relation to APCs such as DCs and macrophages (135) we hypothesized that co-stimulatory molecules might trigger the suppressive effect of Tregs on osteoclasts. Our observation that both genetic and pharmacological neutralization of CD80/86 completely restored osteoclast formation in the presence of Tregs, clearly pointed to an essential regulatory role of this co-stimulation molecule in osteoclast formation. The physiological importance of this interaction is underlined by the osteoporotic phenotype of $CD80/86^{-/-}$ mice, which lack the regulatory effect of Tregs on osteoclast differentiation as the ligand to succeed suppression is missing on osteoclast precursor cells. This explains the enhanced osteoclast formation and bone resorption we observed *in vivo*. This effect is not redundant among other co-stimulation molecules, but specific for CD80/86, as the deletion of ICOS, ICOSL as well as CD28 did not affect the suppressive effect of Tregs on osteoclasts and did not result in a similar bone phenotype *in vivo*. The slight but not significant decrease in systemic bone mass in $CD28^{-/-}$ could be explained by the reduced numbers of Tregs in these mice as co-stimulation via CD28 is needed to induce expression of *Foxp3* (136).

The central regulatory role of CD80/86 in osteoclast formation appears to be linked to the immunomodulatory enzyme indoleamine 2,3-dioxygenase (IDO). By constitutive expression of CTLA-4 on Tregs, these cells can bind to CD80/86 on DCs and induce IDO expression (137). IDO is a potent regulatory molecule that induces the catabolism of tryptophan into pro-apoptotic metabolites such as kynurenine resulting in decreased potential to activate effector T cells (77). Other reported regulatory mechanisms of IDO upregulation are, the prevention of T cell clonal expansion and blocking of the ability of T cells to respond to DCs (137). In accordance, activation of IDO and tryptophan degradation in APCs was shown to induce T cell tolerance and leads to apoptosis of T cells and monocytes, which suggests that IDO expressing APCs, especially macrophages and DCs, may represent a discrete subset of professional APCs (108). Moreover, the presence of IL-10, which is also secreted by Tregs, resulted in sustained expression of functional IDO in mature DCs (108). Interestingly, we observed that exposure of osteoclast precursors to Tregs or to CTLA-4 Ig induced IDO expression and activity, which was evident by accumulation of kynurenine, the major product of tryptophan catabolism. This induction of IDO in osteoclast precursors enhanced apoptosis and thereby inhibited further differentiation into mature osteoclasts. More importantly, functional osteoclastogenesis despite presence of Tregs or CTLA-4 Ig could be restored when IDO was down-regulated by specific siRNA. These data indicate that the regulatory effect of IDO is not only confined to the communication between cells of the innate immune system and T cells but also extends to the osteoclast lineage. Modulation of IDO activity in these cells by Tregs, influences bone homeostasis and tightly links immune regulation with regulation of bone mass.

Our data highlight the importance of tryptophan metabolism in bone remodeling and nicely connect its already known role in bone formation to bone resorption. Regarding the role of tryptophan metabolism in bone formation an important finding in the field of bone biology was the link between low-density lipoprotein receptor (LDLR)-related protein 5 (Lrp5), and bone mass in humans and in mice. Loss of function in Lrp5 leads to the osteoporosis pseudo-glioma syndrome (OPPG), with low bone mass, whereas gain of function leads to the high bone mass (HBM) phenotype in humans (138-140). It was shown that Lrp5 acts through inhibition of tryptophan hydroxylase 1 (Tph1) expression in enterochromaffin cells (141). Thp1 is the rate-limiting enzyme in serotonin synthesis out of L-tryptophan, and serotonin is known for its inhibitory effects on osteoblast proliferation that results in higher bone mass (142). Taken together, our findings show that (i) tryptophan catabolism into serotonin leads to a decrease in osteoblast proliferation and bone formation but on the other hand, (ii) tryptophan catabolism into kynurenine in osteoclast precursors leads decreased osteoclast differentiation and bone resorption.

4.3. Tregs in bone diseases and normal bone homeostasis

On the basis of our results so far we continued investigating the suppressive effect of Tregs on osteoclast differentiation and bone homeostasis *in vivo,* both under pathological and normal conditions. Rheumatoid arthritis (RA) is one of the most common human autoimmune diseases, with prevalence of nearly 1%. Treatment of RA patients with TNF blockers results in a significant clinical benefit (143). Although the frequencies of Tregs in the synovial fluids of RA patients were enhanced compared to the peripheral blood (124) and blood of healthy donors, recent studies in RA patients have suggested that the function of Tregs may be impaired. They are still able to suppress proliferation of $CD4^+CD25^-$ T cells, (122) but are unable to suppress the secretion of pro-inflammatory cytokines such as IFNγ (123). Ehrenstein et al. have initially shown that anti-TNFα therapy in RA patients restored the full suppressive capacity of Tregs on $CD4^+CD25^-$ T cell proliferation as well as on cytokine secretion (123). However, 3 years later the same group published new data that led to the reinterpretation of former results. They have shown that anti-TNFα therapy did not restore the suppressive phenotype of defective Tregs in RA patients but give rise to a new, TGFβ induced, subpopulation of Tregs ($CD4^+CD25^+CD62L^-$) (144). In addition, exposure to TNFα up-regulated expression of TNF-Receptor II (TNFRII) on Tregs, and binding of TNF to TNFRII reversed their suppressive activity by down-regulating *FOXP3* (145). In another recent study, it was demonstrated that defects in Treg function in RA patients result from defects in CTLA-4 surface expression (Flores-Borja F., PNAS, 2008). The authors found reduced expression and functional abnormalities of CTLA-4 on Tregs from RA patients. Furthermore they did provide evidence that induced expression of CTLA-4 on Tregs from RA patients *in vitro* restored their suppressive capacity (146). Also, a study about genetic risk factors in RA patients linked a CTLA-4 polymorphisms with RA susceptibility (147). These studies are in line with our *in vitro* and *in vivo* results, identifying CTLA-4 as important ligand for cell-contact dependent suppression.

None of the studies published so far paid attention to the possibility that Tregs could, in addition to suppress T cell responses, also directly suppress osteoclast differentiation. The importance of Tregs in RA was so far highlighted by studies investigating their role in suppressing inflammation in the collagen-induced arthritis (CIA) mouse model. CIA is induced through immunization with bovine type II collagen (CII) emulsified in Freund`s complete adjuvant and is a commonly used model to study RA (148). It was shown that depletion of $CD4^+CD25^+$ regulatory T cells accelerates the onset of CIA in mice (149) and that CIA could be effectively treated by adoptive transfer of $CD4^+CD25^+$ regulatory T cells (127). Moreover, the defective suppressive potential of $CD4^+CD25^+$ regulatory T cells in CIA mice could partially be restored by IFNγ (150). In our study, we also observed a protective effect of Tregs in our *TNF*tg arthritis model, namely on (i) inflammation, (ii) inflammation induced local bone loss and most importantly (iii) systemic bone loss. Protection from systemic bone loss is uncoupled from local inflammation and indicates direct suppressive effect on osteoclast differentiation by Tregs.

In line with the rescue of the arthritic phenotype in h*TNF*tg mice after bone marrow transfer from *foxp3*tg mice, we observed similar effects in *hTNF*tg mice after treatment with superagonistic CD28-specific monoclonal antibodies (supCD28Mab). It was demonstrated that supCD28Mab are capable of expanding functional $CD4^+CD25^+$ Treg numbers (151), which was highly effective in the treatment of experimental autoimmune encephalomyelitis (107). The potential of the supCD28Mab to selectively increase Treg numbers in *hTNF*tg mice was sufficient to ameliorate the clinical and histological signs of inflammation induced arthritis in these mice. Moreover, similar to the data we obtained from *foxp3*tg *hTNF*tg bone marrow chimera animals, treatment also resulted in an increase of systemic bone mass and a decrease in osteoclast numbers.

As a second bone disease we investigated the role of Tregs in postmenopausal osteoporosis. Osteoporosis is a condition characterized by the loss of normal bone density, resulting in fragile bone. It is most common in women after menopause (postmenopausal bone loss). Hormonal factors strongly determine the rate of bone resorption in postmenopausal bone loss, especially the lack of estrogens. Estrogen deficiency can be mimicked by taking away the main source of estrogens, the ovaries, from mice (ovx mice). This is a generally accepted mouse model for osteoporosis and was used in the present study as second model to investigate the role of Tregs in pathological bone conditions. T cells were shown to play a central role in ovx-induced osteoporosis since estrogen deficiency leads to an increased production of TNFα by T cells (152). TNFα in turn enhances osteoclast formation by increasing production of M-CSF and RANKL by stromal cell (50, 153) and increases the sensitivity of osteoclast precursors to RANKL (105, 152). Ovariectomy did not change the level of TNF secreted by T cells, but instead increases the number of TNF producing T cells (154). This increased T cell proliferation is a result of enhanced T cell activation by antigen presenting bone marrow macrophages and DCs that have up-regulated their expression of MHCII due to the loss of estrogen (155). In addition, it was shown that T cell deficient athymic nude mice (*nu/nu*) mice were fully protected against ovx-induced bone loss (152). In our ovariectomy model, we observed a protective effect due to increased numbers of Tregs in *foxp3*tg mice, which were resistant to ovx- induced on systemic bone loss. This effect correlates well with a lack of increase in numbers of osteoclasts in *foxp3*tg mice. In accordance to that, we detected low CTX-I serum levels, as marker for bone resorption and an unchanged pool of osteoblasts and osteocalcin levels, as markers for bone formation. These results most likely pointed to a decrease in bone resorption, as the main mechanism of how Treg protects from ovx- induced for bone loss.

In both of these models, however, we cannot exclude the possibility that Tregs also suppress the proliferation of TNF secreting T cells and therefore indirectly reduce the number of bone resorbing osteoclast. Finally, to investigate the role of Tregs under normal, healthy conditions we analyzed mice with an increased Treg cell compartment as well as wild-type mice after adoptive Tregs transfer in order to visualize their effect on bone homeostasis. It was shown that scurfin, the protein product of *Foxp3*, is directly involved in generating Tregs; mice over-expressing *Foxp3* (*foxp3*tg) show an increased percentage (15-20%) of Tregs compared to non-transgenic littermate controls (control) (7-10%) (62, 153). So far, *Foxp3* still represents the most Treg specific gene identified for naturally occurring Tregs (62, 101, 104). For these reasons, we have analyzed *foxp3*tg mice in order study the role of Tregs in healthy bone homeostasis. Interestingly we found that the number of osteoclasts *in vivo* as well as CTX-I serum levels, were consistently blunted in *foxp3*tg mice similar to our observation in RAG1 deficient mice following adoptive Tregs transfer. Moreover, no change in osteoblast number, mineral apposition rate or osteocalcin levels in the serum were observed in *foxp3*tg mice or RAG1 deficient mice following adoptive Treg transfer. These data and the fact that Tregs were conspicuously localized along trabecular bone clearly suggest that the bone-homeostatic role of Foxp3$^+$ Tregs acts locally and is mediated through osteoclasts. One possibility how Foxp3 in *foxp3*tg mice affects osteoclast formation could be an intrinsically impairment of osteoclast formation. This assumption is highly unlikely since monocytes of wild-type and *foxp3*tg mice did not express Foxp3, which is rather confined to the T cell lineage (156). In addition, direct osteoclast differentiation assays without T cells did not show any difference in the osteoclastogenic differentiation potential among monocytes from wild-type and *foxp3*tg mice.

Taking together our results from the *in vivo* studies allow three potential mechanisms how Tregs could principally affect osteoclast differentiation *in vivo*: (i) The RANKL/OPG ratio could be shifted in favour of OPG with Tregs, which impairs RANKL to engage RANK and thereby prevent osteoclast formation. This has been ruled out by previously showing an even higher expression of RANKL in Tregs compared to activated T cells (99). In addition, no clear change in the circulating

OPG/RANKL ratio could be detected in the blood of *foxp3*tg mice or in Tregs-transferred RAG1 deficient mice. (ii) Tregs could shift the cytokine milieu from pro-osteoclastogenic cytokines such as TNFα, IL-1, IL-6 or IL-17 to an anti-osteoclastogenic pattern with more, IL-4, IL-10 or IFNγ. Our co-culture experiments of Tregs and osteoclasts did not favour this hypothesis, since blockade of IL-10 or IL-4 did only slightly impair the effect of Tregs on osteoclasts (99). Kim and Kelchtermanns who independently reported an *in vitro* suppressive effect of Tregs on human and mouse osteoclast differentiation, emphasized the role of soluble cytokines as source of suppression (97, 98). Zauli et al. and others (157-159) showed that TRAIL, a TNF superfamily member might be a potential link between the three studies (97-99). It is likely that production of membrane-bound as well as soluble TRAIL by Tregs could indeed influence the formation of osteoclasts and at least partly explain the regulatory effect on osteoclast differentiation (160) and thereby explain the differences observed by us and others. Our results show a mild but significant increase in the level of circulating IFNγ in *foxp3*tg mice and we found increased IFNγ in the supernatant of co-cultures of Tregs with osteoclast precursors. Thus, a role for secreted cytokines, mainly IFNγ in Tregs inhibition of osteoclastogenesis can not be fully excluded. (iii) Tregs could inhibit osteoclastogenesis by suppressing activated T cells. This hypothesis cannot be excluded in the case of local bone loss in the context of arthritis, where bone is directly exposed to inflammatory tissue. However, since also systemic bone in our arthritis model, which is not directly exposed to inflammatory tissue, is well protected by Tregs and since this bone-sparing effect extends to ovariectomy-induced bone loss and physiologically increased bone mass of *foxp3*tg mice, this concept is unlikely. Most importantly, however, is the fact that Tregs can inhibit osteoclastogenesis in T cell deficient mice, which strongly argues in favour of a direct effect of Tregs on bone without involvement of other T cells. Thus, the most attractive concept is that Tregs directly impair osteoclast differentiation via direct cell-contact in vivo. The observation that Foxp3$^+$ Tregs accumulate at sites of intense bone remodeling such as the epiphysis of long bones and sites of bone destruction in case of arthritis, also favours the concept that Tregs need direct cell-contact and act locally on bone.

5. Concluding Remarks

Our conclusions that naturally occurring regulatory T cells (Tregs) suppress osteoclast differentiation *in vitro* and *in vivo* are based on the following experimental observations. (I) Tregs suppress osteoclast differentiation *in vitro* in a cell-contact dependent manner. Cytokines expressed by Tregs have an additional suppressive effect but are not essential. (II) Bone mass is increased in mice with increased Treg function. (III) These mice are also partially protected against osteoporosis following estrogen depletion. (IV) While bone is more heavily destroyed in mice with no Tregs, increased Treg numbers are protective. (V) We observed a decreased bone mass in $CD80/86^{-/-}$ mice as the ligand for CTLA-4 on osteoclast precursors is missing whereas other co-stimulatory mutant mice showed no similar bone phenotype. (VI) Increased bone mass is seen in T cell deficient mice following adoptive transfer of Tregs.

In summary, our data open a new perspective in the interaction between the immune system and the skeleton. The central concept of osteoimmunology that inflammatory T cell subsets such as activated Th1 cells and Th17 cells drive bone loss is now enriched by the concept that immune homeostasis is tightly linked to bone homeostasis. Thus, a key mechanism of immune regulation, the naturally occurring Foxp3$^+$ regulatory T cells, transduces its favourable role in immune balance to skeletal homeostasis. This principle appears conceivable, because activation of the immune system and activation of bone resorption may go hand in hand. This may drive new concepts of a "bone protection system" and shape immunologic tools to maintain bone mass.

References

6. References

1. Kong, Y.Y., Yoshida, H., Sarosi, I., Tan, H.L., Timms, E., Capparelli, C., Morony, S., Oliveira-dos-Santos, A.J., Van, G., Itie, A., et al. 1999. OPGL is a key regulator of osteoclastogenesis, lymphocyte development and lymph-node organogenesis. *Nature* 397:315-323.
2. Lacey, D.L., Timms, E., Tan, H.L., Kelley, M.J., Dunstan, C.R., Burgess, T., Elliott, R., Colombero, A., Elliott, G., Scully, S., et al. 1998. Osteoprotegerin ligand is a cytokine that regulates osteoclast differentiation and activation. *Cell* 93:165-176.
3. Horton, M.A., and Davies, J. 1989. Perspectives: adhesion receptors in bone. *J Bone Miner Res* 4:803-808.
4. Soriano, P., Montgomery, C., Geske, R., and Bradley, A. 1991. Targeted disruption of the c-src proto-oncogene leads to osteopetrosis in mice. *Cell* 64:693-702.
5. Tanaka, S., Takahashi, N., Udagawa, N., Tamura, T., Akatsu, T., Stanley, E.R., Kurokawa, T., and Suda, T. 1993. Macrophage colony-stimulating factor is indispensable for both proliferation and differentiation of osteoclast progenitors. *J Clin Invest* 91:257-263.
6. Yao, Z., Li, P., Zhang, Q., Schwarz, E.M., Keng, P., Arbini, A., Boyce, B.F., and Xing, L. 2006. Tumor necrosis factor-alpha increases circulating osteoclast precursor numbers by promoting their proliferation and differentiation in the bone marrow through up-regulation of c-Fms expression. *J Biol Chem* 281:11846-11855.
7. Yoshida, H., Hayashi, S., Kunisada, T., Ogawa, M., Nishikawa, S., Okamura, H., Sudo, T., and Shultz, L.D. 1990. The murine mutation osteopetrosis is in the coding region of the macrophage colony stimulating factor gene. *Nature* 345:442-444.
8. Dai, X.M., Ryan, G.R., Hapel, A.J., Dominguez, M.G., Russell, R.G., Kapp, S., Sylvestre, V., and Stanley, E.R. 2002. Targeted disruption of the mouse colony-stimulating factor 1 receptor gene results in osteopetrosis, mononuclear phagocyte deficiency, increased primitive progenitor cell frequencies, and reproductive defects. *Blood* 99:111-120.

9. Burgess, T.L., Qian, Y., Kaufman, S., Ring, B.D., Van, G., Capparelli, C., Kelley, M., Hsu, H., Boyle, W.J., Dunstan, C.R., et al. 1999. The ligand for osteoprotegerin (OPGL) directly activates mature osteoclasts. *J Cell Biol* 145:527-538.

10. Bucay, N., Sarosi, I., Dunstan, C.R., Morony, S., Tarpley, J., Capparelli, C., Scully, S., Tan, H.L., Xu, W., Lacey, D.L., et al. 1998. osteoprotegerin-deficient mice develop early onset osteoporosis and arterial calcification. *Genes Dev* 12:1260-1268.

11. Mizuno, A., Amizuka, N., Irie, K., Murakami, A., Fujise, N., Kanno, T., Sato, Y., Nakagawa, N., Yasuda, H., Mochizuki, S., et al. 1998. Severe osteoporosis in mice lacking osteoclastogenesis inhibitory factor/osteoprotegerin. *Biochem Biophys Res Commun* 247:610-615.

12. Simonet, W.S., Lacey, D.L., Dunstan, C.R., Kelley, M., Chang, M.S., Luthy, R., Nguyen, H.Q., Wooden, S., Bennett, L., Boone, T., et al. 1997. Osteoprotegerin: a novel secreted protein involved in the regulation of bone density. *Cell* 89:309-319.

13. Kim, N., Odgren, P.R., Kim, D.K., Marks, S.C., Jr., and Choi, Y. 2000. Diverse roles of the tumor necrosis factor family member TRANCE in skeletal physiology revealed by TRANCE deficiency and partial rescue by a lymphocyte-expressed TRANCE transgene. *Proc Natl Acad Sci U S A* 97:10905-10910.

14. Li, J., Sarosi, I., Yan, X.Q., Morony, S., Capparelli, C., Tan, H.L., McCabe, S., Elliott, R., Scully, S., Van, G., et al. 2000. RANK is the intrinsic hematopoietic cell surface receptor that controls osteoclastogenesis and regulation of bone mass and calcium metabolism. *Proc Natl Acad Sci U S A* 97:1566-1571.

15. Dougall, W.C., Glaccum, M., Charrier, K., Rohrbach, K., Brasel, K., De Smedt, T., Daro, E., Smith, J., Tometsko, M.E., Maliszewski, C.R., et al. 1999. RANK is essential for osteoclast and lymph node development. *Genes Dev* 13:2412-2424.

16. Lomaga, M.A., Yeh, W.C., Sarosi, I., Duncan, G.S., Furlonger, C., Ho, A., Morony, S., Capparelli, C., Van, G., Kaufman, S., et al. 1999. TRAF6 deficiency results in osteopetrosis and defective interleukin-1, CD40, and LPS signaling. *Genes Dev* 13:1015-1024.

17. David, J.P., Sabapathy, K., Hoffmann, O., Idarraga, M.H., and Wagner, E.F. 2002. JNK1 modulates osteoclastogenesis through both c-Jun phosphorylation-dependent and -independent mechanisms. *J Cell Sci* 115:4317-4325.
18. Franzoso, G., Carlson, L., Xing, L., Poljak, L., Shores, E.W., Brown, K.D., Leonardi, A., Tran, T., Boyce, B.F., and Siebenlist, U. 1997. Requirement for NF-kappaB in osteoclast and B-cell development. *Genes Dev* 11:3482-3496.
19. Grigoriadis, A.E., Wang, Z.Q., Cecchini, M.G., Hofstetter, W., Felix, R., Fleisch, H.A., and Wagner, E.F. 1994. c-Fos: a key regulator of osteoclast-macrophage lineage determination and bone remodeling. *Science* 266:443-448.
20. Ducy, P., Zhang, R., Geoffroy, V., Ridall, A.L., and Karsenty, G. 1997. Osf2/Cbfa1: a transcriptional activator of osteoblast differentiation. *Cell* 89:747-754.
21. Ducy, P., Desbois, C., Boyce, B., Pinero, G., Story, B., Dunstan, C., Smith, E., Bonadio, J., Goldstein, S., Gundberg, C., et al. 1996. Increased bone formation in osteocalcin-deficient mice. *Nature* 382:448-452.
22. Komori, T., Yagi, H., Nomura, S., Yamaguchi, A., Sasaki, K., Deguchi, K., Shimizu, Y., Bronson, R.T., Gao, Y.H., Inada, M., et al. 1997. Targeted disruption of Cbfa1 results in a complete lack of bone formation owing to maturational arrest of osteoblasts. *Cell* 89:755-764.
23. Otto, F., Thornell, A.P., Crompton, T., Denzel, A., Gilmour, K.C., Rosewell, I.R., Stamp, G.W., Beddington, R.S., Mundlos, S., Olsen, B.R., et al. 1997. Cbfa1, a candidate gene for cleidocranial dysplasia syndrome, is essential for osteoblast differentiation and bone development. *Cell* 89:765-771.
24. Lanyon, L.E. 1993. Osteocytes, strain detection, bone modeling and remodeling. *Calcif Tissue Int* 53 Suppl 1:S102-106; discussion S106-107.
25. Naski, M.C., Wang, Q., Xu, J., and Ornitz, D.M. 1996. Graded activation of fibroblast growth factor receptor 3 by mutations causing achondroplasia and thanatophoric dysplasia. *Nat Genet* 13:233-237.
26. Moon, R.T., Kohn, A.D., De Ferrari, G.V., and Kaykas, A. 2004. WNT and beta-catenin signalling: diseases and therapies. *Nat Rev Genet* 5:691-701.

27. Yang, Y., Topol, L., Lee, H., and Wu, J. 2003. Wnt5a and Wnt5b exhibit distinct activities in coordinating chondrocyte proliferation and differentiation. *Development* 130:1003-1015.
28. Mackie, E.J., Ahmed, Y.A., Tatarczuch, L., Chen, K.S., and Mirams, M. 2008. Endochondral ossification: how cartilage is converted into bone in the developing skeleton. *Int J Biochem Cell Biol* 40:46-62.
29. Zelzer, E., Glotzer, D.J., Hartmann, C., Thomas, D., Fukai, N., Soker, S., and Olsen, B.R. 2001. Tissue specific regulation of VEGF expression during bone development requires Cbfa1/Runx2. *Mech Dev* 106:97-106.
30. Zelzer, E., Mamluk, R., Ferrara, N., Johnson, R.S., Schipani, E., and Olsen, B.R. 2004. VEGFA is necessary for chondrocyte survival during bone development. *Development* 131:2161-2171.
31. Kim, C.H., Takai, E., Zhou, H., von Stechow, D., Muller, R., Dempster, D.W., and Guo, X.E. 2003. Trabecular bone response to mechanical and parathyroid hormone stimulation: the role of mechanical microenvironment. *J Bone Miner Res* 18:2116-2125.
32. Chapuy, M.C., Arlot, M.E., Duboeuf, F., Brun, J., Crouzet, B., Arnaud, S., Delmas, P.D., and Meunier, P.J. 1992. Vitamin D3 and calcium to prevent hip fractures in the elderly women. *N Engl J Med* 327:1637-1642.
33. Hollis, B.W. 2005. Circulating 25-hydroxyvitamin D levels indicative of vitamin D sufficiency: implications for establishing a new effective dietary intake recommendation for vitamin D. *J Nutr* 135:317-322.
34. Wang, J., Zhou, J., Cheng, C.M., Kopchick, J.J., and Bondy, C.A. 2004. Evidence supporting dual, IGF-I-independent and IGF-I-dependent, roles for GH in promoting longitudinal bone growth. *J Endocrinol* 180:247-255.
35. Weinstein, R.S., Jilka, R.L., Parfitt, A.M., and Manolagas, S.C. 1998. Inhibition of osteoblastogenesis and promotion of apoptosis of osteoblasts and osteocytes by glucocorticoids. Potential mechanisms of their deleterious effects on bone. *J Clin Invest* 102:274-282.
36. Kameda, T., Mano, H., Yuasa, T., Mori, Y., Miyazawa, K., Shiokawa, M., Nakamaru, Y., Hiroi, E., Hiura, K., Kameda, A., et al. 1997. Estrogen inhibits bone resorption by directly inducing apoptosis of the bone-resorbing osteoclasts. *J Exp Med* 186:489-495.

37. Srivastava, S., Toraldo, G., Weitzmann, M.N., Cenci, S., Ross, F.P., and Pacifici, R. 2001. Estrogen decreases osteoclast formation by down-regulating receptor activator of NF-kappa B ligand (RANKL)-induced JNK activation. *J Biol Chem* 276:8836-8840.
38. Glass, D.A., 2nd, and Karsenty, G. 2007. In vivo analyses of Wnt signaling in bone. *Endocrinology* 148:2630-2634.
39. Bafico, A., Liu, G., Yaniv, A., Gazit, A., and Aaronson, S.A. 2001. Novel mechanism of Wnt signalling inhibition mediated by Dickkopf-1 interaction with LRP6/Arrow. *Nat Cell Biol* 3:683-686.
40. Glinka, A., Wu, W., Delius, H., Monaghan, A.P., Blumenstock, C., and Niehrs, C. 1998. Dickkopf-1 is a member of a new family of secreted proteins and functions in head induction. *Nature* 391:357-362.
41. Morvan, F., Boulukos, K., Clement-Lacroix, P., Roman Roman, S., Suc-Royer, I., Vayssiere, B., Ammann, P., Martin, P., Pinho, S., Pognonec, P., et al. 2006. Deletion of a single allele of the Dkk1 gene leads to an increase in bone formation and bone mass. *J Bone Miner Res* 21:934-945.
42. Diarra, D., Stolina, M., Polzer, K., Zwerina, J., Ominsky, M.S., Dwyer, D., Korb, A., Smolen, J., Hoffmann, M., Scheinecker, C., et al. 2007. Dickkopf-1 is a master regulator of joint remodeling. *Nat Med* 13:156-163.
43. Hsu, H., Lacey, D.L., Dunstan, C.R., Solovyev, I., Colombero, A., Timms, E., Tan, H.L., Elliott, G., Kelley, M.J., Sarosi, I., et al. 1999. Tumor necrosis factor receptor family member RANK mediates osteoclast differentiation and activation induced by osteoprotegerin ligand. *Proc Natl Acad Sci U S A* 96:3540-3545.
44. Li, X., Qin, L., Bergenstock, M., Bevelock, L.M., Novack, D.V., and Partridge, N.C. 2007. Parathyroid hormone stimulates osteoblastic expression of MCP-1 to recruit and increase the fusion of pre/osteoclasts. *J Biol Chem* 282:33098-33106.
45. Kim, M.S., Day, C.J., Selinger, C.I., Magno, C.L., Stephens, S.R., and Morrison, N.A. 2006. MCP-1-induced human osteoclast-like cells are tartrate-resistant acid phosphatase, NFATc1, and calcitonin receptor-positive but require receptor activator of NFkappaB ligand for bone resorption. *J Biol Chem* 281:1274-1285.

46. Zhao, C., Irie, N., Takada, Y., Shimoda, K., Miyamoto, T., Nishiwaki, T., Suda, T., and Matsuo, K. 2006. Bidirectional ephrinB2-EphB4 signaling controls bone homeostasis. *Cell Metab* 4:111-121.
47. Mundy, G.R., and Elefteriou, F. 2006. Boning up on ephrin signaling. *Cell* 126:441-443.
48. Moonga, B.S., Adebanjo, O.A., Wang, H.J., Li, S., Wu, X.B., Troen, B., Inzerillo, A., Abe, E., Minkin, C., Huang, C.L., et al. 2002. Differential effects of interleukin-6 receptor activation on intracellular signaling and bone resorption by isolated rat osteoclasts. *J Endocrinol* 173:395-405.
49. Sims, N.A., Jenkins, B.J., Quinn, J.M., Nakamura, A., Glatt, M., Gillespie, M.T., Ernst, M., and Martin, T.J. 2004. Glycoprotein 130 regulates bone turnover and bone size by distinct downstream signaling pathways. *J Clin Invest* 113:379-389.
50. Hofbauer, L.C., Lacey, D.L., Dunstan, C.R., Spelsberg, T.C., Riggs, B.L., and Khosla, S. 1999. Interleukin-1beta and tumor necrosis factor-alpha, but not interleukin-6, stimulate osteoprotegerin ligand gene expression in human osteoblastic cells. *Bone* 25:255-259.
51. Allman, D., Sambandam, A., Kim, S., Miller, J.P., Pagan, A., Well, D., Meraz, A., and Bhandoola, A. 2003. Thymopoiesis independent of common lymphoid progenitors. *Nat Immunol* 4:168-174.
52. Godfrey, V., Wilkinson, J.E., and Russell, L.B. 1991. X-linked lymphoreticular disease in the scurfy (*sf*) mutant. *Am. J. Pathol.* 138:1379-1387.
53. Griesenbach, U., Boyton, R.J., Somerton, L., Garcia, S.E., Ferrari, S., Owaki, T., Ya-Fen, Z., Geddes, D.M., Hasegawa, M., Altmann, D.M., et al. 2006. Effect of tolerance induction to immunodominant T-cell epitopes of Sendai virus on gene expression following repeat administration to lung. *Gene Ther* 13:449-456.
54. Mosmann, T.R., Cherwinski, H., Bond, M.W., Giedlin, M.A., and Coffman, R.L. 1986. Two types of murine helper T cell clone. I. Definition according to profiles of lymphokine activities and secreted proteins. *J Immunol* 136:2348-2357.
55. Romagnani, S. 2006. Regulation of the T cell response. *Clin Exp Allergy* 36:1357-1366.

56. Zhou, L., Chong, M.M., and Littman, D.R. 2009. Plasticity of CD4+ T cell lineage differentiation. *Immunity* 30:646-655.
57. Nakayama, T., and Yamashita, M. 2008. Initiation and maintenance of Th2 cell identity. *Curr Opin Immunol* 20:265-271.
58. Veldhoen, M., Hocking, R.J., Atkins, C.J., Locksley, R.M., and Stockinger, B. 2006. TGFbeta in the context of an inflammatory cytokine milieu supports de novo differentiation of IL-17-producing T cells. *Immunity* 24:179-189.
59. Weaver, C.T., Hatton, R.D., Mangan, P.R., and Harrington, L.E. 2007. IL-17 family cytokines and the expanding diversity of effector T cell lineages. *Annu Rev Immunol* 25:821-852.
60. Sakaguchi, S., Sakaguchi, N., Asano, M., Itoh, M., and Toda, M. 1995. Immunologic self-tolerance maintained by activated T cells expressing IL-2 receptor alpha-chains (CD25). Breakdown of a single mechanism of self-tolerance causes various autoimmune diseases. *J Immunol* 155:1151-1164.
61. Bennett, C.L., Christie, J., Ramsdell, F., Brunkow, M.E., Ferguson, P.J., Whitesell, L., Kelly, T.E., Saulsbury, F.T., Chance, P.F., and Ochs, H.D. 2001. The immune dysregulation, polyendocrinopathy, enteropathy, X-linked syndrome (IPEX) is caused by mutations of FOXP3. *Nat Genet* 27:20-21.
62. Brunkow, M.E., Jeffery, E.W., Hjerrild, K.A., Paeper, B., Clark, L.B., Yasayko, S.A., Wilkinson, J.E., Galas, D., Ziegler, S.F., and Ramsdell, F. 2001. Disruption of a new forkhead/winged-helix protein, scurfin, results in the fatal lymphoproliferative disorder of the scurfy mouse. *Nat Genet* 27:68-73.
63. Chatila, T.A., Blaeser, F., Ho, N., Lederman, H.M., Voulgaropoulos, C., Helms, C., and Bowcock, A.M. 2000. JM2, encoding a fork head-related protein, is mutated in X-linked autoimmunity-allergic disregulation syndrome. *J Clin Invest* 106:R75-81.
64. Wildin, R.S., Ramsdell, F., Peake, J., Faravelli, F., Casanova, J.L., Buist, N., Levy-Lahad, E., Mazzella, M., Goulet, O., Perroni, L., et al. 2001. X-linked neonatal diabetes mellitus, enteropathy and endocrinopathy syndrome is the human equivalent of mouse scurfy. *Nat Genet* 27:18-20.
65. Fontenot, J.D., and Rudensky, A.Y. 2005. A well adapted regulatory contrivance: regulatory T cell development and the forkhead family transcription factor Foxp3. *Nat Immunol* 6:331-337.

66. Hori, S., Nomura, T., and Sakaguchi, S. 2003. Control of regulatory T cell development by the transcription factor Foxp3. *Science* 299:1057-1061.
67. Gavin, M.A., Rasmussen, J.P., Fontenot, J.D., Vasta, V., Manganiello, V.C., Beavo, J.A., and Rudensky, A.Y. 2007. Foxp3-dependent programme of regulatory T-cell differentiation. *Nature* 445:771-775.
68. Wan, Y.Y., and Flavell, R.A. 2007. Regulatory T cells, transforming growth factor-beta, and immune suppression. *Proc Am Thorac Soc* 4:271-276.
69. Williams, L.M., and Rudensky, A.Y. 2007. Maintenance of the Foxp3-dependent developmental program in mature regulatory T cells requires continued expression of Foxp3. *Nat Immunol* 8:277-284.
70. Takahashi, T., Kuniyasu, Y., Toda, M., Sakaguchi, N., Itoh, M., Iwata, M., Shimizu, J., and Sakaguchi, S. 1998. Immunologic self-tolerance maintained by CD25+CD4+ naturally anergic and suppressive T cells: induction of autoimmune disease by breaking their anergic/suppressive state. *Int Immunol* 10:1969-1980.
71. Thornton, A.M., and Shevach, E.M. 1998. CD4+CD25+ immunoregulatory T cells suppress polyclonal T cell activation in vitro by inhibiting interleukin 2 production. *J Exp Med* 188:287-296.
72. Pandiyan, P., Zheng, L., Ishihara, S., Reed, J., and Lenardo, M.J. 2007. CD4+CD25+Foxp3+ regulatory T cells induce cytokine deprivation-mediated apoptosis of effector CD4+ T cells. *Nat Immunol* 8:1353-1362.
73. Collison, L.W., Workman, C.J., Kuo, T.T., Boyd, K., Wang, Y., Vignali, K.M., Cross, R., Sehy, D., Blumberg, R.S., and Vignali, D.A. 2007. The inhibitory cytokine IL-35 contributes to regulatory T-cell function. *Nature* 450:566-569.
74. Garin, M.I., Chu, C.C., Golshayan, D., Cernuda-Morollon, E., Wait, R., and Lechler, R.I. 2007. Galectin-1: a key effector of regulation mediated by CD4+CD25+ T cells. *Blood* 109:2058-2065.
75. Wing, K., Onishi, Y., Prieto-Martin, P., Yamaguchi, T., Miyara, M., Fehervari, Z., Nomura, T., and Sakaguchi, S. 2008. CTLA-4 control over Foxp3+ regulatory T cell function. *Science* 322:271-275.
76. Serra, P., Amrani, A., Yamanouchi, J., Han, B., Thiessen, S., Utsugi, T., Verdaguer, J., and Santamaria, P. 2003. CD40 ligation releases immature dendritic cells from the control of regulatory CD4+CD25+ T cells. *Immunity* 19:877-889.

77. Grohmann, U., Orabona, C., Fallarino, F., Vacca, C., Calcinaro, F., Falorni, A., Candeloro, P., Belladonna, M.L., Bianchi, R., Fioretti, M.C., et al. 2002. CTLA-4-Ig regulates tryptophan catabolism in vivo. *Nat Immunol* 3:1097-1101.
78. Calvi, L.M., Adams, G.B., Weibrecht, K.W., Weber, J.M., Olson, D.P., Knight, M.C., Martin, R.P., Schipani, E., Divieti, P., Bringhurst, F.R., et al. 2003. Osteoblastic cells regulate the haematopoietic stem cell niche. *Nature* 425:841-846.
79. Zhang, J., Niu, C., Ye, L., Huang, H., He, X., Tong, W.G., Ross, J., Haug, J., Johnson, T., Feng, J.Q., et al. 2003. Identification of the haematopoietic stem cell niche and control of the niche size. *Nature* 425:836-841.
80. Mazo, I.B., Honczarenko, M., Leung, H., Cavanagh, L.L., Bonasio, R., Weninger, W., Engelke, K., Xia, L., McEver, R.P., Koni, P.A., et al. 2005. Bone marrow is a major reservoir and site of recruitment for central memory CD8+ T cells. *Immunity* 22:259-270.
81. Gilbert, L., He, X., Farmer, P., Boden, S., Kozlowski, M., Rubin, J., and Nanes, M.S. 2000. Inhibition of osteoblast differentiation by tumor necrosis factor-alpha. *Endocrinology* 141:3956-3964.
82. Lind, M., Deleuran, B., Yssel, H., Fink-Eriksen, E., and Thestrup-Pedersen, K. 1995. IL-4 and IL-13, but not IL-10, are chemotactic factors for human osteoblasts. *Cytokine* 7:78-82.
83. Onoe, Y., Miyaura, C., Kaminakayashiki, T., Nagai, Y., Noguchi, K., Chen, Q.R., Seo, H., Ohta, H., Nozawa, S., Kudo, I., et al. 1996. IL-13 and IL-4 inhibit bone resorption by suppressing cyclooxygenase-2-dependent prostaglandin synthesis in osteoblasts. *J Immunol* 156:758-764.
84. Horton, J.E., Raisz, L.G., Simmons, H.A., Oppenheim, J.J., and Mergenhagen, S.E. 1972. Bone resorbing activity in supernatant fluid from cultured human peripheral blood leukocytes. *Science* 177:793-795.
85. Dewhirst, F.E., Stashenko, P.P., Mole, J.E., and Tsurumachi, T. 1985. Purification and partial sequence of human osteoclast-activating factor: identity with interleukin 1 beta. *J Immunol* 135:2562-2568.
86. Kong, Y.Y., Feige, U., Sarosi, I., Bolon, B., Tafuri, A., Morony, S., Capparelli, C., Li, J., Elliott, R., McCabe, S., et al. 1999. Activated T cells regulate bone loss and joint destruction in adjuvant arthritis through osteoprotegerin ligand. *Nature* 402:304-309.

87. Horwood, N.J., Kartsogiannis, V., Quinn, J.M., Romas, E., Martin, T.J., and Gillespie, M.T. 1999. Activated T lymphocytes support osteoclast formation in vitro. *Biochem Biophys Res Commun* 265:144-150.
88. Osman, N., Ley, S.C., and Crumpton, M.J. 1992. Evidence for an association between the T cell receptor/CD3 antigen complex and the CD5 antigen in human T lymphocytes. *Eur J Immunol* 22:2995-3000.
89. Takayanagi, H., Ogasawara, K., Hida, S., Chiba, T., Murata, S., Sato, K., Takaoka, A., Yokochi, T., Oda, H., Tanaka, K., et al. 2000. T-cell-mediated regulation of osteoclastogenesis by signalling cross-talk between RANKL and IFN-gamma. *Nature* 408:600-605.
90. Choi, Y., Woo, K.M., Ko, S.H., Lee, Y.J., Park, S.J., Kim, H.M., and Kwon, B.S. 2001. Osteoclastogenesis is enhanced by activated B cells but suppressed by activated CD8(+) T cells. *Eur J Immunol* 31:2179-2188.
91. Dai, S.M., Matsuno, H., Nakamura, H., Nishioka, K., and Yudoh, K. 2004. Interleukin-18 enhances monocyte tumor necrosis factor alpha and interleukin-1beta production induced by direct contact with T lymphocytes: implications in rheumatoid arthritis. *Arthritis Rheum* 50:432-443.
92. Sato, K., Suematsu, A., Okamoto, K., Yamaguchi, A., Morishita, Y., Kadono, Y., Tanaka, S., Kodama, T., Akira, S., Iwakura, Y., et al. 2006. Th17 functions as an osteoclastogenic helper T cell subset that links T cell activation and bone destruction. *J Exp Med* 203:2673-2682.
93. Kotake, S., Udagawa, N., Takahashi, N., Matsuzaki, K., Itoh, K., Ishiyama, S., Saito, S., Inoue, K., Kamatani, N., Gillespie, M.T., et al. 1999. IL-17 in synovial fluids from patients with rheumatoid arthritis is a potent stimulator of osteoclastogenesis. *J Clin Invest* 103:1345-1352.
94. Choi, S.J., Cruz, J.C., Craig, F., Chung, H., Devlin, R.D., Roodman, G.D., and Alsina, M. 2000. Macrophage inflammatory protein 1-alpha is a potential osteoclast stimulatory factor in multiple myeloma. *Blood* 96:671-675.
95. Scheven, B.A., Milne, J.S., Hunter, I., and Robins, S.P. 1999. Macrophage-inflammatory protein-1alpha regulates preosteoclast differentiation in vitro. *Biochem Biophys Res Commun* 254:773-778.
96. Li, Y., Toraldo, G., Li, A., Yang, X., Zhang, H., Qian, W.P., and Weitzmann, M.N. 2007. B cells and T cells are critical for the preservation of bone homeostasis and attainment of peak bone mass in vivo. *Blood* 109:3839-3848.

97. Kelchtermans, H., Geboes, L., Mitera, T., Huskens, D., Leclercq, G., and Matthys, P. 2009. Activated CD4+CD25+ regulatory T cells inhibit osteoclastogenesis and collagen-induced arthritis. *Ann Rheum Dis* 68:744-750.
98. Kim, Y.G., Lee, C.K., Nah, S.S., Mun, S.H., Yoo, B., and Moon, H.B. 2007. Human CD4+CD25+ regulatory T cells inhibit the differentiation of osteoclasts from peripheral blood mononuclear cells. *Biochem Biophys Res Commun* 357:1046-1052.
99. Zaiss, M.M., Axmann, R., Zwerina, J., Polzer, K., Guckel, E., Skapenko, A., Schulze-Koops, H., Horwood, N., Cope, A., and Schett, G. 2007. Treg cells suppress osteoclast formation: a new link between the immune system and bone. *Arthritis Rheum* 56:4104-4112.
100. Keffer, J., Probert, L., Cazlaris, H., Georgopoulos, S., Kaslaris, E., Kioussis, D., and Kollias, G. 1991. Transgenic mice expressing human tumour necrosis factor: a predictive genetic model of arthritis. *The EMBO Journal* 10:4025-4031.
101. Khattri, R., Kasprowicz, D., Cox, T., Mortrud, M., Appleby, M.W., Brunkow, M.E., Ziegler, S.F., and Ramsdell, F. 2001. The amount of scurfin protein determines peripheral T cell number and responsiveness. *J Immunol* 167:6312-6320.
102. McAdam, A.J., Greenwald, R.J., Levin, M.A., Chernova, T., Malenkovich, N., Ling, V., Freeman, G.J., and Sharpe, A.H. 2001. ICOS is critical for CD40-mediated antibody class switching. *Nature* 409:102-105.
103. Shahinian, A., Pfeffer, K., Lee, K.P., Kundig, T.M., Kishihara, K., Wakeham, A., Kawai, K., Ohashi, P.S., Thompson, C.B., and Mak, T.W. 1993. Differential T cell costimulatory requirements in CD28-deficient mice. *Science* 261:609-612.
104. Khattri, R., Cox, T., Yasayko, S.A., and Ramsdell, F. 2003. An essential role for Scurfin in CD4+CD25+ T regulatory cells. *Nat Immunol* 4:337-342.
105. Lam, J., Takeshita, S., Barker, J.E., Kanagawa, O., Ross, F.P., and Teitelbaum, S.L. 2000. TNF-alpha induces osteoclastogenesis by direct stimulation of macrophages exposed to permissive levels of RANK ligand. *J Clin Invest* 106:1481-1488.

106. Schett, G., Redlich, K., Hayer, S., Zwerina, J., Bolon, B., Dunstan, C., Gortz, B., Schulz, A., Bergmeister, H., Kollias, G., et al. 2003. Osteoprotegerin protects against generalized bone loss in tumor necrosis factor-transgenic mice. *Arthritis Rheum* 48:2042-2051.

107. Beyersdorf, N., Gaupp, S., Balbach, K., Schmidt, J., Toyka, K.V., Lin, C.H., Hanke, T., Hunig, T., Kerkau, T., and Gold, R. 2005. Selective targeting of regulatory T cells with CD28 superagonists allows effective therapy of experimental autoimmune encephalomyelitis. *J Exp Med* 202:445-455.

108. Munn, D.H., Sharma, M.D., Lee, J.R., Jhaver, K.G., Johnson, T.S., Keskin, D.B., Marshall, B., Chandler, P., Antonia, S.J., Burgess, R., et al. 2002. Potential regulatory function of human dendritic cells expressing indoleamine 2,3-dioxygenase. *Science* 297:1867-1870.

109. Goldring, S.R. 2003. Inflammatory mediators as essential elements in bone remodeling. *Calcif Tissue Int* 73:97-100.

110. Siwik, D.A., Chang, D.L., and Colucci, W.S. 2000. Interleukin-1beta and tumor necrosis factor-alpha decrease collagen synthesis and increase matrix metalloproteinase activity in cardiac fibroblasts in vitro. *Circ Res* 86:1259-1265.

111. Horwood, N.J., Elliott, J., Martin, T.J., and Gillespie, M.T. 1998. Osteotropic agents regulate the expression of osteoclast differentiation factor and osteoprotegerin in osteoblastic stromal cells. *Endocrinology* 139:4743-4746.

112. Kitazawa, R., Kimble, R.B., Vannice, J.L., Kung, V.T., and Pacifici, R. 1994. Interleukin-1 receptor antagonist and tumor necrosis factor binding protein decrease osteoclast formation and bone resorption in ovariectomized mice. *J Clin Invest* 94:2397-2406.

113. Thomson, B.M., Mundy, G.R., and Chambers, T.J. 1987. Tumor necrosis factors alpha and beta induce osteoblastic cells to stimulate osteoclastic bone resorption. *J Immunol* 138:775-779.

114. Li, P., Schwarz, E.M., O'Keefe, R.J., Ma, L., Boyce, B.F., and Xing, L. 2004. RANK signaling is not required for TNFalpha-mediated increase in CD11(hi) osteoclast precursors but is essential for mature osteoclast formation in TNFalpha-mediated inflammatory arthritis. *J Bone Miner Res* 19:207-213.

115. Wei, S., Wang, M.W., Teitelbaum, S.L., and Ross, F.P. 2002. Interleukin-4 reversibly inhibits osteoclastogenesis via inhibition of NF-kappa B and mitogen-activated protein kinase signaling. *J Biol Chem* 277:6622-6630.
116. Sato, K., and Takayanagi, H. 2006. Osteoclasts, rheumatoid arthritis, and osteoimmunology. *Curr Opin Rheumatol* 18:419-426.
117. Theill, L.E., Boyle, W.J., and Penninger, J.M. 2002. RANK-L and RANK: T cells, bone loss, and mammalian evolution. *Annu Rev Immunol* 20:795-823.
118. Kelsen, J., Agnholt, J., Hoffmann, H.J., Romer, J.L., Hvas, C.L., and Dahlerup, J.F. 2005. FoxP3(+)CD4(+)CD25(+) T cells with regulatory properties can be cultured from colonic mucosa of patients with Crohn's disease. *Clin Exp Immunol* 141:549-557.
119. Crispin, J.C., Martinez, A., and Alcocer-Varela, J. 2003. Quantification of regulatory T cells in patients with systemic lupus erythematosus. *J Autoimmun* 21:273-276.
120. Kukreja, A., Cost, G., Marker, J., Zhang, C., Sun, Z., Lin-Su, K., Ten, S., Sanz, M., Exley, M., Wilson, B., et al. 2002. Multiple immuno-regulatory defects in type-1 diabetes. *J Clin Invest* 109:131-140.
121. Longhi, M.S., Hussain, M.J., Mitry, R.R., Arora, S.K., Mieli-Vergani, G., Vergani, D., and Ma, Y. 2006. Functional study of CD4+CD25+ regulatory T cells in health and autoimmune hepatitis. *J Immunol* 176:4484-4491.
122. Cao, D., van Vollenhoven, R., Klareskog, L., Trollmo, C., and Malmstrom, V. 2004. CD25brightCD4+ regulatory T cells are enriched in inflamed joints of patients with chronic rheumatic disease. *Arthritis Res Ther* 6:R335-346.
123. Ehrenstein, M.R., Evans, J.G., Singh, A., Moore, S., Warnes, G., Isenberg, D.A., and Mauri, C. 2004. Compromised function of regulatory T cells in rheumatoid arthritis and reversal by anti-TNFalpha therapy. *J Exp Med* 200:277-285.
124. van Amelsfort, J.M., Jacobs, K.M., Bijlsma, J.W., Lafeber, F.P., and Taams, L.S. 2004. CD4(+)CD25(+) regulatory T cells in rheumatoid arthritis: differences in the presence, phenotype, and function between peripheral blood and synovial fluid. *Arthritis Rheum* 50:2775-2785.
125. Waterhouse, P., Penninger, J.M., Timms, E., Wakeham, A., Shahinian, A., Lee, K.P., Thompson, C.B., Griesser, H., and Mak, T.W. 1995.

Lymphoproliferative disorders with early lethality in mice deficient in Ctla-4. *Science* 270:985-988.

126. Shull, M.M., Ormsby, I., Kier, A.B., Pawlowski, S., Diebold, R.J., Yin, M., Allen, R., Sidman, C., Proetzel, G., Calvin, D., et al. 1992. Targeted disruption of the mouse transforming growth factor-beta 1 gene results in multifocal inflammatory disease. *Nature* 359:693-699.

127. Morgan, M.E., Flierman, R., van Duivenvoorde, L.M., Witteveen, H.J., van Ewijk, W., van Laar, J.M., de Vries, R.R., and Toes, R.E. 2005. Effective treatment of collagen-induced arthritis by adoptive transfer of CD25+ regulatory T cells. *Arthritis Rheum* 52:2212-2221.

128. Nguyen, L.T., Jacobs, J., Mathis, D., and Benoist, C. 2007. Where FoxP3-dependent regulatory T cells impinge on the development of inflammatory arthritis. *Arthritis Rheum* 56:509-520.

129. Misra, N., Bayry, J., Lacroix-Desmazes, S., Kazatchkine, M.D., and Kaveri, S.V. 2004. Cutting edge: human CD4+CD25+ T cells restrain the maturation and antigen-presenting function of dendritic cells. *J Immunol* 172:4676-4680.

130. Paust, S., and Cantor, H. 2005. Regulatory T cells and autoimmune disease. *Immunol Rev* 204:195-207.

131. Hase, H., Kanno, Y., Kojima, H., Sakurai, D., and Kobata, T. 2008. Co-culture of osteoclast precursors with rheumatoid synovial fibroblasts induces osteoclastogenesis via transforming growth factor beta-mediated down-regulation of osteoprotegerin. *Arthritis Rheum* 58:3356-3365.

132. Stein, N.C., Kreutzmann, C., Zimmermann, S.P., Niebergall, U., Hellmeyer, L., Goettsch, C., Schoppet, M., and Hofbauer, L.C. 2008. Interleukin-4 and interleukin-13 stimulate the osteoclast inhibitor osteoprotegerin by human endothelial cells through the STAT6 pathway. *J Bone Miner Res* 23:750-758.

133. Mohamed, S.G., Sugiyama, E., Shinoda, K., Taki, H., Hounoki, H., Abdel-Aziz, H.O., Maruyama, M., Kobayashi, M., Ogawa, H., and Miyahara, T. 2007. Interleukin-10 inhibits RANKL-mediated expression of NFATc1 in part via suppression of c-Fos and c-Jun in RAW264.7 cells and mouse bone marrow cells. *Bone* 41:592-602.

134. Axmann, R., Herman, S., Zaiss, M., Franz, S., Polzer, K., Zwerina, J., Herrmann, M., Smolen, J., and Schett, G. 2008. CTLA-4 directly inhibits osteoclast formation. *Ann Rheum Dis* 67:1603-1609.

135. Alnaeeli, M., Park, J., Mahamed, D., Penninger, J.M., and Teng, Y.T. 2007. Dendritic cells at the osteo-immune interface: implications for inflammation-induced bone loss. *J Bone Miner Res* 22:775-780.

136. Tai, X., Cowan, M., Feigenbaum, L., and Singer, A. 2005. CD28 costimulation of developing thymocytes induces Foxp3 expression and regulatory T cell differentiation independently of interleukin 2. *Nat Immunol* 6:152-162.

137. Mellor, A.L., Chandler, P., Baban, B., Hansen, A.M., Marshall, B., Pihkala, J., Waldmann, H., Cobbold, S., Adams, E., and Munn, D.H. 2004. Specific subsets of murine dendritic cells acquire potent T cell regulatory functions following CTLA4-mediated induction of indoleamine 2,3 dioxygenase. *Int Immunol* 16:1391-1401.

138. Boyden, L.M., Mao, J., Belsky, J., Mitzner, L., Farhi, A., Mitnick, M.A., Wu, D., Insogna, K., and Lifton, R.P. 2002. High bone density due to a mutation in LDL-receptor-related protein 5. *N Engl J Med* 346:1513-1521.

139. Gong, Y., Slee, R.B., Fukai, N., Rawadi, G., Roman-Roman, S., Reginato, A.M., Wang, H., Cundy, T., Glorieux, F.H., Lev, D., et al. 2001. LDL receptor-related protein 5 (LRP5) affects bone accrual and eye development. *Cell* 107:513-523.

140. Little, R.D., Carulli, J.P., Del Mastro, R.G., Dupuis, J., Osborne, M., Folz, C., Manning, S.P., Swain, P.M., Zhao, S.C., Eustace, B., et al. 2002. A mutation in the LDL receptor-related protein 5 gene results in the autosomal dominant high-bone-mass trait. *Am J Hum Genet* 70:11-19.

141. Yadav, V.K., Ryu, J.H., Suda, N., Tanaka, K.F., Gingrich, J.A., Schutz, G., Glorieux, F.H., Chiang, C.Y., Zajac, J.D., Insogna, K.L., et al. 2008. Lrp5 controls bone formation by inhibiting serotonin synthesis in the duodenum. *Cell* 135:825-837.

142. Gustafsson, B.I., Westbroek, I., Waarsing, J.H., Waldum, H., Solligard, E., Brunsvik, A., Dimmen, S., van Leeuwen, J.P., Weinans, H., and Syversen, U. 2006. Long-term serotonin administration leads to higher bone mineral density, affects bone architecture, and leads to higher femoral bone stiffness in rats. *J Cell Biochem* 97:1283-1291.

143. Lipsky, P.E., van der Heijde, D.M., St Clair, E.W., Furst, D.E., Breedveld, F.C., Kalden, J.R., Smolen, J.S., Weisman, M., Emery, P., Feldmann, M., et al. 2000. Infliximab and methotrexate in the treatment of rheumatoid arthritis.

Anti-Tumor Necrosis Factor Trial in Rheumatoid Arthritis with Concomitant Therapy Study Group. *N Engl J Med* 343:1594-1602.

144. Nadkarni, S., Mauri, C., and Ehrenstein, M.R. 2007. Anti-TNF-alpha therapy induces a distinct regulatory T cell population in patients with rheumatoid arthritis via TGF-beta. *J Exp Med* 204:33-39.

145. Valencia, X., Stephens, G., Goldbach-Mansky, R., Wilson, M., Shevach, E.M., and Lipsky, P.E. 2006. TNF downmodulates the function of human CD4+CD25hi T-regulatory cells. *Blood* 108:253-261.

146. Flores-Borja, F., Jury, E.C., Mauri, C., and Ehrenstein, M.R. 2008. Defects in CTLA-4 are associated with abnormal regulatory T cell function in rheumatoid arthritis. *Proc Natl Acad Sci U S A* 105:19396-19401.

147. Walker, E.J., Hirschfield, G.M., Xu, C., Lu, Y., Liu, X., Coltescu, C., Wang, K., Newman, W.G., Bykerk, V., Keystone, E.C., et al. 2009. CTLA4/ICOS gene variants and haplotypes are associated with rheumatoid arthritis and primary biliary cirrhosis in the Canadian population. *Arthritis Rheum* 60:931-937.

148. Anthony, D.D., and Haqqi, T.M. 1999. Collagen-induced arthritis in mice: an animal model to study the pathogenesis of rheumatoid arthritis. *Clin Exp Rheumatol* 17:240-244.

149. Morgan, M.E., Sutmuller, R.P., Witteveen, H.J., van Duivenvoorde, L.M., Zanelli, E., Melief, C.J., Snijders, A., Offringa, R., de Vries, R.R., and Toes, R.E. 2003. CD25+ cell depletion hastens the onset of severe disease in collagen-induced arthritis. *Arthritis Rheum* 48:1452-1460.

150. Kelchtermans, H., De Klerck, B., Mitera, T., Van Balen, M., Bullens, D., Billiau, A., Leclercq, G., and Matthys, P. 2005. Defective CD4+CD25+ regulatory T cell functioning in collagen-induced arthritis: an important factor in pathogenesis, counter-regulated by endogenous IFN-gamma. *Arthritis Res Ther* 7:R402-415.

151. Lin, C.H., and Hunig, T. 2003. Efficient expansion of regulatory T cells in vitro and in vivo with a CD28 superagonist. *Eur J Immunol* 33:626-638.

152. Cenci, S., Weitzmann, M.N., Roggia, C., Namba, N., Novack, D., Woodring, J., and Pacifici, R. 2000. Estrogen deficiency induces bone loss by enhancing T-cell production of TNF-alpha. *J Clin Invest* 106:1229-1237.

153. Sherman, M.L., Weber, B.L., Datta, R., and Kufe, D.W. 1990. Transcriptional and posttranscriptional regulation of macrophage-specific colony stimulating

factor gene expression by tumor necrosis factor. Involvement of arachidonic acid metabolites. *J Clin Invest* 85:442-447.

154. Roggia, C., Gao, Y., Cenci, S., Weitzmann, M.N., Toraldo, G., Isaia, G., and Pacifici, R. 2001. Up-regulation of TNF-producing T cells in the bone marrow: a key mechanism by which estrogen deficiency induces bone loss in vivo. *Proc Natl Acad Sci U S A* 98:13960-13965.

155. Cenci, S., Toraldo, G., Weitzmann, M.N., Roggia, C., Gao, Y., Qian, W.P., Sierra, O., and Pacifici, R. 2003. Estrogen deficiency induces bone loss by increasing T cell proliferation and lifespan through IFN-gamma-induced class II transactivator. *Proc Natl Acad Sci U S A* 100:10405-10410.

156. Ziegler, S.F. 2006. FOXP3: of mice and men. *Annu Rev Immunol* 24:209-226.

157. Roux, S., Lambert-Comeau, P., Saint-Pierre, C., Lepine, M., Sawan, B., and Parent, J.L. 2005. Death receptors, Fas and TRAIL receptors, are involved in human osteoclast apoptosis. *Biochem Biophys Res Commun* 333:42-50.

158. Zauli, G., Rimondi, E., Nicolin, V., Melloni, E., Celeghini, C., and Secchiero, P. 2004. TNF-related apoptosis-inducing ligand (TRAIL) blocks osteoclastic differentiation induced by RANKL plus M-CSF. *Blood* 104:2044-2050.

159. Zauli, G., Rimondi, E., Stea, S., Baruffaldi, F., Stebel, M., Zerbinati, C., Corallini, F., and Secchiero, P. 2008. TRAIL inhibits osteoclastic differentiation by counteracting RANKL-dependent p27Kip1 accumulation in pre-osteoclast precursors. *J Cell Physiol* 214:117-125.

160. Zaiss, M.M., and Schett, G. 2008. Reply Letter. *Arthritis & Rheumatism* 58:1887-1888.

I want morebooks!

Buy your books fast and straightforward online - at one of the world's fastest growing online book stores! Environmentally sound due to Print-on-Demand technologies.

Buy your books online at
www.get-morebooks.com

Kaufen Sie Ihre Bücher schnell und unkompliziert online – auf einer der am schnellsten wachsenden Buchhandelsplattformen weltweit! Dank Print-On-Demand umwelt- und ressourcenschonend produziert.

Bücher schneller online kaufen
www.morebooks.de

OmniScriptum Marketing DEU GmbH
Heinrich-Böcking-Str. 6-8
D - 66121 Saarbrücken
Telefax: +49 681 93 81 567-9

info@omniscriptum.com
www.omniscriptum.com

Printed by Books on Demand GmbH, Norderstedt / Germany